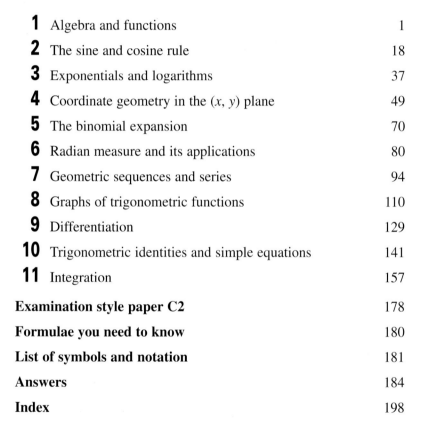

HEINEMANN MODULAR MATHEMATICS
for
EDEXCEL AS AND A-LEVEL
Pure Mathematics C2

Greg Attwood   Alistair Macpherson   Bronwen Moran
Joe Petran   Keith Pledger   Geoff Staley   Dave Wilkins

Endorsed by edexcel

27146A

heinemann.co.uk
✓ Free online support
✓ Useful weblinks
✓ 24 hour online ordering

01865 888058

Heinemann
Inspiring generations

Heinemann Educational Publishers
Halley Court, Jordan Hill, Oxford OX2 8EJ
Part of Harcourt Education

Heinemann is the registered trademark of
Harcourt Education Limited

© Greg Attwood, Alistair David Macpherson, Bronwen Moran, Joe Petran, Keith Pledger, Geoff Staley,
Dave Wilkins 2004

First published 2004

09 08 07 06 05 04
10 9 8 7 6 5 4 3 2 1

British Library Cataloguing in Publication Data is available
from the British Library on request.

ISBN 0 435 51098 3

Edited by Richard Beatty
Designed by Bridge Creative Services
Typeset by Tech-Set Ltd

Original illustrations © Harcourt Education Limited, 2004

Illustrated by Tech-Set Ltd

Cover design by Bridge Creation Services

Printed in Spain by Mateu Cromo S.A.

Acknowledgements
Every effort has been made to contact copyright holders of material reproduced in this book. Any omissions will
be rectified in subsequent printings if notice is given to the publishers.

Tel: 01865 888058 email: info.he@heinemann.co.uk

Contents

About this book

This book is designed to provide you with the best preparation possible for your Edexcel C2 exam. The authors are members of a senior examining team themselves and have a good understanding of Edexcel's requirements.

Finding your way round

To help you find your way around when you are studying and revising use the:

- **edge colours** – each chapter has a different colour scheme. This helps you to get to the right chapter quickly.
- **contents list** – this lists the headings that identify key syllabus ideas covered in the book so you can turn straight to them. The detailed contents list shows which parts of the C2 syllabus are covered in each section.
- **index** – this lists the headings that identify key syllabus ideas covered in this book so you can turn straight to them.

How sections are structured

- Each section (e.g. 1.1, 1.2) begins with a statement. The statement tells you what is covered in the section.

1.3 You can expand an expression by multiplying each term inside the bracket by the term outside.

- Some sections include explanations, which help you understand the maths behind the questions you need to answer in your exam.
- Examples are worked through step-by-step. They are model solutions, as you might write them out. Examiners' hints are given in yellow margin note boxes.
- Each section ends with an exercise, with plenty of questions for practice.

Remembering key ideas

Key ideas you need to remember are listed in a summary of key points at the end of each chapter. The statement at the beginning of a section may be a key point. When key points appear in the teaching, they are marked like this:

■ **The square root of a prime number is a surd.**

Exercises and exam questions

In this book questions are carefully graded so they increase in difficulty and gradually bring you up to standard.

- **Past exam questions** are marked with an [E].
- **Mixed exercises** at the end of each chapter help you practice answering questions on all the topics you have covered in the chapter.
- **Exam style practice paper** is designed to help you prepare for the exam itself.
- **Answers** are included at the end of the book – use the answers to check your work.

1 Algebra and functions

This chapter shows you how to do algebraic division and use the remainder and factor theorems.

1.1 You can simplify algebraic fractions by division.

Example 1

Simplify these fractions:

a $\dfrac{7x^4 - 2x^3 + 6x}{x}$ 　　　　 **b** $\dfrac{5x^2 - 6}{2x}$ 　　　　 **c** $\dfrac{3x^5 - 4x^2}{-3x}$

a $\dfrac{7x^4 - 2x^3 + 6x}{x}$

$= \dfrac{7x^4}{x} - \dfrac{2x^3}{x} + \dfrac{6x}{x}$ ————————— Divide each term on top of the fraction by x.

Simplify the individual fractions when possible, so that:

① $\dfrac{7x^4}{x} = 7 \times \dfrac{x^4}{x}$

$= 7 \times x^{4-1}$

$= 7x^3$

② $\dfrac{2x^3}{x} = 2 \times \dfrac{x^3}{x}$

$= 2 \times x^{3-1}$

$= 2x^2$

③ $\dfrac{6x}{x} = 6 \times \dfrac{x}{x}$

$= 6 \times x^{1-1}$

$= 6x^0$

$= 6$ ————————— Remember $x^0 = 1$.

So $\dfrac{7x^4 - 2x^3 + 6x}{x} = 7x^3 - 2x^2 + 6$.

b $\dfrac{5x^2 - 6}{2x}$

$= \dfrac{5x^2}{2x} - \dfrac{6}{2x}$ •————— Divide each term on top of the fraction by $2x$.

Simplify the fractions, so that:

① $\dfrac{5x^2}{2x} = \dfrac{5}{2} \times \dfrac{x^2}{x}$

$= \dfrac{5}{2} \times x^{2-1}$

$= \dfrac{5}{2}x$ •————— Remember $x^1 = x$ and $\dfrac{5}{2}x = \dfrac{5x}{2}$.

② $-\dfrac{6}{2x} = -\dfrac{3}{x}$

So $\dfrac{5x^2 - 6}{2x} = \dfrac{5x}{2} - \dfrac{3}{x}$.

c $\dfrac{3x^5 - 4x^2}{-3x}$

$= \dfrac{3x^5}{-3x} - \dfrac{4x^2}{-3x}$ •————— Divide each term on top of the fraction by $-3x$.

Simplify the fractions, so that:

① $\dfrac{3x^5}{-3x} = \dfrac{3}{-3} \times \dfrac{x^5}{x}$

$= -1 \times x^{5-1}$

$= -x^4$

② $\dfrac{-4x^2}{-3x} = \dfrac{-4}{-3} \times \dfrac{x^2}{x}$

$= \dfrac{4}{3} \times x^{2-1}$ •————— Minus divided by minus equals plus.

$= \dfrac{4}{3}x$

So $\dfrac{3x^5 - 4x^2}{-3x} = -x^4 + \dfrac{4x}{3}$. •————— $\dfrac{4}{3}x = \dfrac{4x}{3}$

Example 2

Simplify these fractions by factorising:

a $\dfrac{(x+7)(2x-1)}{(2x-1)}$ **b** $\dfrac{x^2+7x+12}{(x+3)}$ **c** $\dfrac{x^2+6x+5}{x^2+3x-10}$ **d** $\dfrac{2x^2+11x+12}{(x+3)(x+4)}$

a $\dfrac{(x+7)(2x-1)}{(2x-1)}$ —————— Simplify by dividing the top and the bottom of the fraction by $(2x-1)$.

$= x+7$

b $\dfrac{x^2+7x+12}{(x+3)}$

$= \dfrac{(x+3)(x+4)}{(x+3)}$ ——— Factorise $x^2+7x+12$:
$(+3)\times(+4)=+12$
$(+3)+(+4)=+7$
So $x^2+7x+12=(x+3)(x+4)$.

$= x+4$ —————— Divide top and bottom by $(x+3)$.

c $\dfrac{x^2+6x+5}{x^2+3x-10}$

$= \dfrac{(x+5)(x+1)}{(x+5)(x-2)}$ —— Factorise x^2+6x+5:
$(+5)\times(+1)=+5$
$(+5)+(+1)=+6$
So $x^2+6x+5=(x+5)(x+1)$.

$= \dfrac{x+1}{x-2}$ — Factorise $x^2+3x-10$:
$(+5)\times(-2)=-10$
$(+5)+(-2)=+3$
So $x^2+3x-10=(x+5)(x-2)$.

Divide top and bottom by $(x+5)$.

d Factorise $2x^2+11x+12$:

$2\times+12=+24$

and $(+3)\times(+8)=+24$
$(+3)+(+8)=+11$

So $2x^2+11x+12 = 2x^2+3x+8x+12$
$= x(2x+3)+4(2x+3)$
$= (2x+3)(x+4)$.

So $\dfrac{2x^2+11x+12}{(x+3)(x+4)}$

$= \dfrac{(2x+3)(x+4)}{(x+3)(x+4)}$

$= \dfrac{2x+3}{x+3}$ —————— Divide top and bottom by $(x+4)$.

Exercise 1A

1 Simplify these fractions:

a $\dfrac{4x^4 + 5x^2 - 7x}{x}$

b $\dfrac{7x^8 - 5x^5 + 9x^3 + x^2}{x}$

c $\dfrac{-2x^3 + x}{x}$

d $\dfrac{-x^4 + 4x^2 + 6}{x}$

e $\dfrac{7x^5 - x^3 - 4}{x}$

f $\dfrac{8x^4 - 4x^3 + 6x}{2x}$

g $\dfrac{9x^2 - 12x^3 - 3x}{3x}$

h $\dfrac{8x^5 - 2x^3}{4x}$

i $\dfrac{7x^3 - x^4 - 2}{5x}$

j $\dfrac{-4x^2 + 6x^4 - 2x}{-2x}$

k $\dfrac{-x^8 + 9x^4 + 6}{-2x}$

l $\dfrac{-9x^9 - 6x^4 - 2}{-3x}$

2 Simplify these fractions as far as possible:

a $\dfrac{(x + 3)(x - 2)}{(x - 2)}$

b $\dfrac{(x + 4)(3x - 1)}{(3x - 1)}$

c $\dfrac{(x + 3)^2}{(x + 3)}$

d $\dfrac{x^2 + 10x + 21}{(x + 3)}$

e $\dfrac{x^2 + 9x + 20}{(x + 4)}$

f $\dfrac{x^2 + x - 12}{(x - 3)}$

g $\dfrac{x^2 + x - 20}{x^2 + 2x - 15}$

h $\dfrac{x^2 + 3x + 2}{x^2 + 5x + 4}$

i $\dfrac{x^2 + x - 12}{x^2 - 9x + 18}$

j $\dfrac{2x^2 + 7x + 6}{(x - 5)(x + 2)}$

k $\dfrac{2x^2 + 9x - 18}{(x + 6)(x + 1)}$

l $\dfrac{3x^2 - 7x + 2}{(3x - 1)(x + 2)}$

m $\dfrac{2x^2 + 3x + 1}{x^2 - x - 2}$

n $\dfrac{x^2 + 6x + 8}{3x^2 + 7x + 2}$

o $\dfrac{2x^2 - 5x - 3}{2x^2 - 9x + 9}$

1.2 You can divide a polynomial by $(x \pm p)$.

Example 3

Divide $x^3 + 2x^2 - 17x + 6$ by $(x - 3)$.

①

$$
\begin{array}{r}
x^2 \\
x - 3{\overline{\smash{\big)}\,x^3 + 2x^2 - 17x + 6}} \\
\underline{x^3 - 3x^2} \\
5x^2 - 17x
\end{array}
$$

Start by dividing the first term of the polynomial by x, so that $x^3 \div x = x^2$.

Next multiply $(x - 3)$ by x^2, so that $x^2 \times (x - 3) = x^3 - 3x^2$.

Now subtract, so that $(x^3 + 2x^2) - (x^3 - 3x^2) = 5x^2$.

Finally copy $-17x$.

②

$$
\begin{array}{r}
x^2 + 5x \\
x - 3{\overline{\smash{\big)}\,x^3 + 2x^2 - 17x + 6}} \\
\underline{x^3 - 3x^2} \\
5x^2 - 17x \\
\underline{5x^2 - 15x} \\
-2x + 6
\end{array}
$$

Repeat the method. Divide $5x^2$ by x, so that $5x^2 \div x = 5x$.

Multiply $(x - 3)$ by $5x$, so that $5x \times (x - 3) = 5x^2 - 15x$.

Subtract, so that $(5x^2 - 17x) - (5x^2 - 15x) = -2x$.

Copy 6.

③

$$
\begin{array}{r}
x^2 + 5x - 2 \\
x - 3{\overline{\smash{\big)}\,x^3 + 2x^2 - 17x + 6}} \\
\underline{x^3 - 3x^2} \\
5x^2 - 17x \\
\underline{5x^2 - 15x} \\
-2x + 6 \\
\underline{-2x + 6} \\
0
\end{array}
$$

Repeat the method. Divide $-2x$ by x, so that $-2x \div x = -2$.

Multiply $(x - 3)$ by -2, so that $-2 \times (x - 3) = -2x + 6$.

Subtract, so that $(-2x + 6) - (-2x + 6) = 0$.

No numbers left to copy, so you have finished.

So $x^3 + 2x^2 - 17x + 6 \div (x - 3) =$
$x^2 + 5x - 2$.

This is called the quotient.

Example 4

Divide $6x^3 + 28x^2 - 7x + 15$ by $(x + 5)$.

①

$$\begin{array}{r} 6x^2 \\ x + 5\overline{)6x^3 + 28x^2 - 7x + 15} \\ \underline{6x^3 + 30x^2} \\ -2x^2 - 7x \end{array}$$

Start by dividing the first term of the polynomial by x, so that $6x^3 \div x = 6x^2$.

Next multiply $(x + 5)$ by $6x^2$, so that $6x^2 \times (x + 5) = 6x^3 + 30x^2$.

Now subtract, so that $(6x^3 + 28x^2) - (6x^3 + 30x^2) = -2x^2$.

Finally copy $-7x$.

②

$$\begin{array}{r} 6x^2 - 2x \\ x + 5\overline{)6x^3 + 28x^2 - 7x + 15} \\ \underline{6x^3 + 30x^2} \\ -2x^2 - 7x \\ \underline{-2x^2 - 10x} \\ 3x + 15 \end{array}$$

Repeat the method. Divide $-2x^2$ by x, so that $-2x^2 \div x = -2x$.

Multiply $(x + 5)$ by $-2x$, so that $-2x \times (x + 5) = -2x^2 - 10x$.

Subtract, so that $(-2x^2 - 7x) - (-2x^2 - 10x) = 3x$.

Copy 15.

③

$$\begin{array}{r} 6x^2 - 2x + 3 \\ x + 5\overline{)6x^3 + 28x^2 - 7x + 15} \\ \underline{6x^3 + 30x^2} \\ -2x^2 - 7x \\ \underline{-2x^2 - 10x} \\ 3x + 15 \\ \underline{3x + 15} \\ 0 \end{array}$$

Repeat the method. Divide $3x$ by x, so that $3x \div x = 3$.

Multiply $(x + 5)$ by 3, so that $3 \times (x + 5) = 3x + 15$.

Subtract, so that $(3x + 15) - (3x + 15) = 0$.

So $6x^3 + 28x^2 - 7x + 15 \div (x + 5) =$
$6x^2 - 2x + 3$.

Example 5

Divide $-3x^4 + 8x^3 - 8x^2 + 13x - 10$ by $(x - 2)$.

①

$$-3x^3$$

$$x - 2\overline{)-3x^4 + 8x^3 - 8x^2 + 13x - 10}$$

$$\underline{-3x^4 + 6x^3}$$

$$2x^3 - 8x^2$$

Start by dividing the first term of the polynomial by x, so that $-3x^4 \div x = -3x^3$.

Next multiply $(x - 2)$ by $-3x^3$, so that $-3x^3 \times (x - 2) = -3x^4 + 6x^3$.

Now subtract, so that $(-3x^4 + 8x^3) - (-3x^4 + 6x^3) = 2x^3$.

Copy $-8x^2$.

②

$$-3x^3 + 2x^2$$

$$x - 2\overline{)-3x^4 + 8x^3 - 8x^2 + 13x - 10}$$

$$\underline{-3x^4 + 6x^3}$$

$$2x^3 - 8x^2$$

$$\underline{2x^3 - 4x^2}$$

$$-4x^2 + 13x$$

Repeat the method. Divide $2x^3$ by x, so that $2x^3 \div x = 2x^2$.

Multiply $(x - 2)$ by $2x^2$, so that $2x^2 \times (x - 2) = 2x^3 - 4x^2$.

Subtract, so that $(2x^3 - 8x^2) - (2x^3 - 4x^2) = -4x^2$.

Copy $13x$.

③

$$-3x^3 + 2x^2 - 4x$$

$$x - 2\overline{)-3x^4 + 8x^3 - 8x^2 + 13x - 10}$$

$$\underline{-3x^4 + 6x^3}$$

$$2x^3 - 8x^2$$

$$\underline{2x^3 - 4x^2}$$

$$-4x^2 + 13x$$

$$\underline{-4x^2 + 8x}$$

$$5x - 10$$

Repeat the method. Divide $-4x^2$ by x, so that $-4x^2 \div x = -4x$.

Multiply $(x - 2)$ by $-4x^2$, so that $-4x^2 \times (x - 2) = -4x^2 + 8x$.

Subtract, so that $(-4x^2 + 13x) - (-4x^2 + 8x) = 5x$.

Copy -10.

④

$$-3x^3 + 2x^2 - 4x + 5$$

$$x - 2\overline{)-3x^4 + 8x^3 - 8x^2 + 13x - 10}$$

$$\underline{-3x^4 + 6x^3}$$

$$2x^3 - 8x^2$$

$$\underline{2x^3 - 4x^2}$$

$$-4x^2 + 13x$$

$$\underline{-4x^2 + 8x}$$

$$5x - 10$$

$$\underline{5x - 10}$$

$$0$$

Repeat the method. Divide $5x$ by x, so that $5x \div x = 5$.

Multiply $(x - 2)$ by 5 so that $5 \times (x - 2) = 5x - 10$.

Subtract, so that $(5x - 10) - (5x - 10) = 0$.

So $-3x^4 + 8x^3 - 8x^2 + 13x - 10 \div (x - 2)$

$= -3x^3 + 2x^2 - 4x + 5$.

Exercise **1B**

1 Divide:
  a $x^3 + 6x^2 + 8x + 3$ by $(x + 1)$
  b $x^3 + 10x^2 + 25x + 4$ by $(x + 4)$
  c $x^3 + 7x^2 - 3x - 54$ by $(x + 6)$
  d $x^3 + 9x^2 + 18x - 10$ by $(x + 5)$
  e $x^3 - x^2 + x + 14$ by $(x + 2)$
  f $x^3 + x^2 - 7x - 15$ by $(x - 3)$
  g $x^3 - 5x^2 + 8x - 4$ by $(x - 2)$
  h $x^3 - 3x^2 + 8x - 6$ by $(x - 1)$
  i $x^3 - 8x^2 + 13x + 10$ by $(x - 5)$
  j $x^3 - 5x^2 - 6x - 56$ by $(x - 7)$

2 Divide:
  a $6x^3 + 27x^2 + 14x + 8$ by $(x + 4)$
  b $4x^3 + 9x^2 - 3x - 10$ by $(x + 2)$
  c $3x^3 - 10x^2 - 10x + 8$ by $(x - 4)$
  d $3x^3 - 5x^2 - 4x - 24$ by $(x - 3)$
  e $2x^3 + 4x^2 - 9x - 9$ by $(x + 3)$
  f $2x^3 - 15x^2 + 14x + 24$ by $(x - 6)$
  g $-3x^3 + 2x^2 - 2x - 7$ by $(x + 1)$
  h $-2x^3 + 5x^2 + 17x - 20$ by $(x - 4)$
  i $-5x^3 - 27x^2 + 23x + 30$ by $(x + 6)$
  j $-4x^3 + 9x^2 - 3x + 2$ by $(x - 2)$

3 Divide:
  a $x^4 + 5x^3 + 2x^2 - 7x + 2$ by $(x + 2)$
  b $x^4 + 11x^3 + 25x^2 - 29x - 20$ by $(x + 5)$
  c $4x^4 + 14x^3 + 3x^2 - 14x - 15$ by $(x + 3)$
  d $3x^4 - 7x^3 - 23x^2 + 14x - 8$ by $(x - 4)$
  e $-3x^4 + 9x^3 - 10x^2 + x + 14$ by $(x - 2)$
  f $3x^5 + 17x^4 + 2x^3 - 38x^2 + 5x - 25$ by $(x + 5)$
  g $6x^5 - 19x^4 + x^3 + x^2 + 13x + 6$ by $(x - 3)$
  h $-5x^5 + 7x^4 + 2x^3 - 7x^2 + 10x - 7$ by $(x - 1)$
  i $2x^6 - 11x^5 + 14x^4 - 16x^3 + 36x^2 - 10x - 24$ by $(x - 4)$
  j $-x^6 + 4x^5 - 4x^4 + 4x^3 - 5x^2 + 7x - 3$ by $(x - 3)$

Example **6**

Divide $x^3 - 3x - 2$ by $(x - 2)$.

$$\begin{array}{r} x^2 + 2x + 1 \\ x - 2 \overline{)x^3 + 0x^2 - 3x - 2} \\ \underline{x^3 - 2x^2} \\ 2x^2 - 3x \\ \underline{2x^2 - 4x} \\ x - 2 \\ \underline{x - 2} \\ 0 \end{array}$$

Use $0x^2$ so that the sum is laid out correctly.

Subtract, so that $(x^3 + 0x^2) - (x^3 - 2x^2) = 2x^2$.

The number remaining after a division is called the **remainder**. In this case the remainder $= 0$, so $(x - 2)$ is a **factor** of $x^3 - 3x - 2$.

So $x^3 - 3x - 2 \div (x - 2) = x^2 + 2x + 1$.

$x^2 + 2x + 1$ is called the **quotient**.

Example 7

Divide $3x^3 - 3x^2 - 4x + 4$ by $(x - 1)$.

$$
\begin{array}{r}
3x^2 \qquad\qquad - 4 \\
x - 1)\overline{3x^3 - 3x^2 - 4x + 4} \\
\underline{3x^3 - 3x^2} \\
\overline{0} - 4x + 4 \\
\underline{-4x + 4} \\
0
\end{array}
$$

Divide $-4x$ by x, so that $-4x \div x = -4$.

Subtract, so that $(3x^3 - 3x^2) - (3x^3 - 3x^2) = 0$.

Copy $-4x$ and 4.

So $x^3 - 3x^2 - 4x + 4 \div (x - 1) =$
$3x^2 - 4$.

Example 8

Find the remainder when $2x^3 - 5x^2 - 16x + 10$ by $(x - 4)$.

$$
\begin{array}{r}
2x^2 + 3x - 4 \\
x - 4)\overline{2x^3 - 5x^2 - 16x + 10} \\
\underline{2x^3 - 8x^2} \\
3x^2 - 16x \\
\underline{3x^2 - 12x} \\
-4x + 10 \\
\underline{-4x + 16} \\
-6
\end{array}
$$

$(x - 4)$ is not a factor of $2x^3 - 5x^2 - 16x + 10$ as the remainder $\neq 0$.

So the remainder is -6.

Exercise 1C

1 Divide:
   a $x^3 + x + 10$ by $(x + 2)$
   b $2x^3 - 17x + 3$ by $(x + 3)$
   c $-3x^3 + 50x - 8$ by $(x - 4)$

2 Divide:
   a $x^3 + x^2 - 36$ by $(x - 3)$
   b $2x^3 + 9x^2 + 25$ by $(x + 5)$
   c $-3x^3 + 11x^2 - 20$ by $(x - 2)$

Hint for question 2:
Use $0x$.

3 Divide:

 a $x^3 + 2x^2 - 5x - 10$ by $(x + 2)$

 b $2x^3 - 6x^2 + 7x - 21$ by $(x - 3)$

 c $-3x^3 + 21x^2 - 4x + 28$ by $(x - 7)$

4 Find the remainder when:

 a $x^3 + 4x^2 - 3x + 2$ is divided by $(x + 5)$

 b $3x^3 - 20x^2 + 10x + 5$ is divided by $(x - 6)$

 c $-2x^3 + 3x^2 + 12x + 20$ is divided by $(x - 4)$

5 Show that when $3x^3 - 2x^2 + 4$ is divided by $(x - 1)$ the remainder is 5.

6 Show that when $3x^4 - 8x^3 + 10x^2 - 3x - 25$ is divided by $(x + 1)$ the remainder is -1.

7 Show that $(x + 4)$ is the factor of $5x^3 - 73x + 28$.

8 Simplify $\dfrac{3x^3 - 8x - 8}{x - 2}$.

> **Hint for question 8:**
> Divide $3x^3 - 8x - 8$ by $(x - 2)$.

9 Divide $x^3 - 1$ by $(x - 1)$.

> **Hint for question 9:**
> Use $0x^2$ and $0x$.

10 Divide $x^4 - 16$ by $(x + 2)$.

1.3 You can factorise a polynomial by using the factor theorem:
If f(x) is a polynomial and f(p) = 0, then $x - p$ is a factor of f(x).

Example **9**

Show that $(x - 2)$ is a factor of $x^3 + x^2 - 4x - 4$ by:
a algebraic division
b the factor theorem

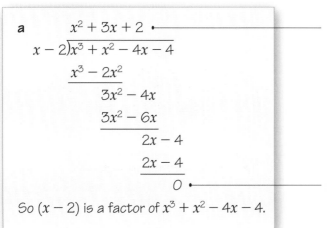

a
$$x^2 + 3x + 2$$
$$x - 2)\overline{x^3 + x^2 - 4x - 4}$$
$$\underline{x^3 - 2x^2}$$
$$3x^2 - 4x$$
$$\underline{3x^2 - 6x}$$
$$2x - 4$$
$$\underline{2x - 4}$$
$$0$$

So $(x - 2)$ is a factor of $x^3 + x^2 - 4x - 4$.

Divide $x^3 + x^2 - 4x - 4$ by $(x - 2)$.

The remainder $= 0$, so $(x - 2)$ is a factor of $x^3 + x^2 - 4x - 4$.

b $f(x) = x^3 + x^2 - 4x - 4$ ——————— Write the polynomial as a function.

$f(2) = (2)^3 + (2)^2 - 4(2) - 4$ ——————— Substitute $x = 2$ into the polynomial.

$= 8 + 4 - 8 - 4$

Use the factor theorem:

$= 0$

If f$(p) = 0$, then $x - p$ is a factor of f(x).

Here $p = 2$, so $(x - 2)$ is a factor of $x^3 + x^2 - 4x - 4$.

So $(x - 2)$ is a factor of $x^3 + x^2 - 4x - 4$.

Example 10

Factorise $2x^3 + x^2 - 18x - 9$.

$f(x) = 2x^3 + x^2 - 18x - 9$ ——————— Write the polynomial as a function.

$f(-1) = 2(-1)^3 + (-1)^2 - 18(-1) - 9 = 8$
$f(1) = 2(1)^3 + (1)^2 - 18(1) - 9 = -24$
$f(2) = 2(2)^3 + (2)^2 - 18(2) - 9 = -25$
$f(3) = 2(3)^3 + (3)^2 - 18(3) - 9 = 0$

Try values of x, e.g. -1, 1, 2, 3, ... until you find f$(p) = 0$.

f$(p) = 0$.

So $(x - 3)$ is a factor of $2x^3 + x^2 - 18x - 9$.

Use the factor theorem:
If f$(p) = 0$, then $x - p$ is a factor of f(x).
Here $p = 3$.

$$\begin{array}{r} 2x^2 + 7x + 3 \\ x - 3\overline{)2x^3 + x^2 - 18x - 9} \\ \underline{2x^3 - 6x^2} \\ 7x^2 - 18x \\ \underline{7x^2 - 21x} \\ 3x - 9 \\ \underline{3x - 9} \\ 0 \end{array}$$

Divide $2x^3 + x^2 - 18x - 9$ by $(x - 3)$.

You can check your division here:
$(x - 3)$ is a factor of $2x^3 + x^2 - 18x - 9$, so the remainder must $= 0$.

$2x^3 + x^2 - 18x - 9 = (x - 3)(2x^2 + 7x + 3)$

$2x^2 + 7x + 3$ can also be factorised.

$= (x - 3)(2x + 1)(x + 3)$

Example 11

Given that $(x + 1)$ is a factor of $4x^4 - 3x^2 + a$, find the value of a.

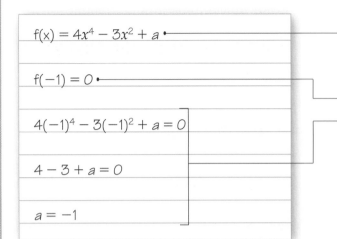

$f(x) = 4x^4 - 3x^2 + a$	Write the polynomial as a function.
$f(-1) = 0$	Use the factor theorem the other way around: $x - p$ is a factor of f(x), so f(p) = 0 Here $p = -1$.
$4(-1)^4 - 3(-1)^2 + a = 0$	Substitute $x = -1$ and solve the equation for a.
$4 - 3 + a = 0$	Remember $(-1)^4 = 1$
$a = -1$	

Example 12

Show that if $(x - p)$ is a factor of f(x) then f(p) = 0.

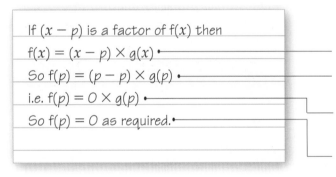

If $(x - p)$ is a factor of $f(x)$ then	
$f(x) = (x - p) \times g(x)$	g(x) is a polynomial expression.
So $f(p) = (p - p) \times g(p)$	Substitute $x = p$.
i.e. $f(p) = 0 \times g(p)$	
So $f(p) = 0$ as required.	$p - p = 0$
	Remember $0 \times$ anything $= 0$.

Exercise 1D

1 Use the factor theorem to show:

    a $(x - 1)$ is a factor of $4x^3 - 3x^2 - 1$

    b $(x + 3)$ is a factor of $5x^4 - 45x^2 - 6x - 18$

    c $(x - 4)$ is a factor of $-3x^3 + 13x^2 - 6x + 8$

2 Show that $(x - 1)$ is a factor of $x^3 + 6x^2 + 5x - 12$ and hence factorise the expression completely.

3 Show that $(x + 1)$ is a factor of $x^3 + 3x^2 - 33x - 35$ and hence factorise the expression completely.

4 Show that $(x - 5)$ is a factor of $x^3 - 7x^2 + 2x + 40$ and hence factorise the expression completely.

5 Show that $(x - 2)$ is a factor of $2x^3 + 3x^2 - 18x + 8$ and hence factorise the expression completely.

6 Each of these expressions has a factor $(x \pm p)$. Find a value of p and hence factorise the expression completely.

    a $x^3 - 10x^2 + 19x + 30$ **b** $x^3 + x^2 - 4x - 4$ **c** $x^3 - 4x^2 - 11x + 30$

7 Factorise:

    a $2x^3 + 5x^2 - 4x - 3$ **b** $2x^3 - 17x^2 + 38x - 15$

    c $3x^3 + 8x^2 + 3x - 2$ **d** $6x^3 + 11x^2 - 3x - 2$

    e $4x^3 - 12x^2 - 7x + 30$

8 Given that $(x - 1)$ is a factor of $5x^3 - 9x^2 + 2x + a$ find the value of a.

9 Given that $(x + 3)$ is a factor of $6x^3 - bx^2 + 18$ find the value of b.

10 Given that $(x - 1)$ and $(x + 1)$ are factors of $px^3 + qx^2 - 3x - 7$ find the value of p and q.

> **Hint for question 10:**
> Solve simultaneous equations.

1.4 You can find the remainder when a polynomial is divided by $(ax - b)$ by using the remainder theorem:
If a polynomial f(x) is divided by $(ax - b)$ then the remainder is $f\left(\dfrac{b}{a}\right)$.

Example **13**

Find the remainder when $x^3 - 20x + 3$ is divided by $(x - 4)$ using:
a algebraic division   **b** the remainder theorem

a

$$\begin{array}{r} x^2 + 4x - 4 \\ x - 4 \overline{)\, x^3 + 0x^2 - 20x + 3\,} \\ x^3 - 4x^2 \\ \hline 4x^2 - 20x \\ 4x^2 - 16x \\ \hline -4x + 3 \\ -4x + 16 \\ \hline -13 \end{array}$$

Divide $x^3 - 20x + 3$ by $(x - 4)$. Remember to use $0x^2$.

The remainder is -13.

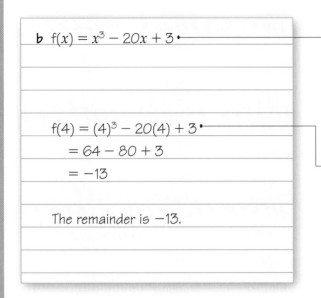

b $f(x) = x^3 - 20x + 3$ •——————————————— Write the polynomial as a function.

Use the remainder theorem: If f(x) is divided by ($ax - b$), then the remainder is $f\left(\dfrac{b}{a}\right)$.

Compare ($x - 4$) to ($ax - b$), so $a = 1$, $b = 4$ and the remainder is $f\left(\dfrac{4}{1}\right)$, i.e. f(4).

$f(4) = (4)^3 - 20(4) + 3$ •

$\quad\quad = 64 - 80 + 3$

$\quad\quad = -13$

Substitute $x = 4$.

The remainder is -13.

Example 14

When $8x^4 - 4x^3 + ax^2 - 1$ is dived by ($2x + 1$) the remainder is 3. Find the value of a.

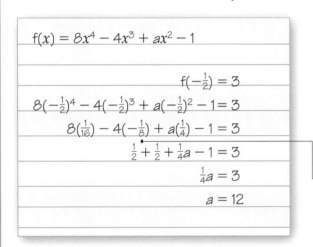

$f(x) = 8x^4 - 4x^3 + ax^2 - 1$

Use the remainder theorem: If f(x) is divided by ($ax - b$), then the remainder is $f\left(\dfrac{b}{a}\right)$.

$f(-\tfrac{1}{2}) = 3$

Compare ($2x + 1$) to ($ax - b$), so $a = 2$, $b = -1$ and the remainder is $f(-\tfrac{1}{2})$, i.e. $f(-\tfrac{1}{2})$.

$8(-\tfrac{1}{2})^4 - 4(-\tfrac{1}{2})^3 + a(-\tfrac{1}{2})^2 - 1 = 3$

$8(\tfrac{1}{16}) - 4(-\tfrac{1}{8}) + a(\tfrac{1}{4}) - 1 = 3$

$\tfrac{1}{2} + \tfrac{1}{2} + \tfrac{1}{4}a - 1 = 3$

Using the fact that the remainder is 3, substitute $x = -\tfrac{1}{2}$ and solve the equation for a.

$\tfrac{1}{4}a = 3$

$a = 12$

$(-\tfrac{1}{2})^3 = -\tfrac{1}{2} \times -\tfrac{1}{2} \times -\tfrac{1}{2} = -\tfrac{1}{8}$

Exercise 1E

1 Find the remainder when:

    a $4x^3 - 5x^2 + 7x + 1$ is divided by ($x - 2$)

    b $2x^5 - 32x^3 + x - 10$ is divided by ($x - 4$)

    c $-2x^3 + 6x^2 + 5x - 3$ is divided by ($x + 1$)

    d $7x^3 + 6x^2 - 45x + 1$ is divided by ($x + 3$)

    e $4x^4 - 4x^2 + 8x - 1$ is divided by ($2x - 1$)

    f $243x^4 - 27x^3 - 3x + 7$ is divided by ($3x - 1$)

    g $64x^3 + 32x^2 - 16x + 9$ is divided by ($4x + 1$)

    h $81x^3 - 81x^2 + 9x + 6$ is divided by ($3x - 2$)

    i $243x^6 - 780x^2 + 6$ is divided by ($3x + 4$)

    j $125x^4 + 5x^3 - 9x$ is divided by ($5x + 3$)

2 When $2x^3 - 3x^2 - 2x + a$ is divided by $(x - 1)$ the remainder is -4. Find the value of a.

3 When $-3x^3 + 4x^2 + bx + 6$ is divided by $(x + 2)$ the remainder is 10. Find the value of b.

4 When $16x^3 - 32x^2 + cx - 8$ is divided by $(2x - 1)$ the remainder is 1. Find the value of c.

5 Show that $(x - 3)$ is a factor of $x^6 - 36x^3 + 243$.

6 Show that $(2x - 1)$ is a factor of $2x^3 + 17x^2 + 31x - 20$.

Hint for question 7:
First find q.

7 $f(x) = x^2 + 3x + q$. Given $f(2) = 3$, find $f(-2)$.

8 $g(x) = x^3 + ax^2 + 3x + 6$. Given $g(-1) = 2$, find the remainder when $g(x)$ is divided by $(3x - 2)$.

9 The expression $2x^3 - x^2 + ax + b$ gives a remainder 14 when divided by $(x - 2)$ and a remainder -86 when divided by $(x + 3)$. Find the value of a and b.

Hint for question 10:
Solve simultaneous equations.

10 The expression $3x^3 + 2x^2 - px + q$ is divisible by $(x - 1)$ but leaves a remainder of 10 when divided by $(x + 1)$. Find the value of a and b.

Mixed exercise 1F

1 Simplify these fractions as far as possible:

a $\dfrac{3x^4 - 21x}{3x}$

b $\dfrac{x^2 - 2x - 24}{x^2 - 7x + 6}$

c $\dfrac{2x^2 + 7x - 4}{2x^2 + 9x + 4}$

2 Divide $3x^3 + 12x^2 + 5x + 20$ by $(x + 4)$.

3 Simplify $\dfrac{2x^3 + 3x + 5}{x + 1}$.

4 Show that $(x - 3)$ is a factor of $2x^3 - 2x^2 - 17x + 15$. Hence express $2x^3 - 2x^2 - 17x + 15$ in the form $(x - 3)(Ax^2 + Bx + C)$, where the values A, B and C are to be found.

5 Show that $(x - 2)$ is a factor of $x^3 + 4x^2 - 3x - 18$. Hence express $x^3 + 4x^2 - 3x - 18$ in the form $(x - 2)(px + q)^2$, where the values p and q are to be found.

6 Factorise completely $2x^3 + 3x^2 - 18x + 8$.

7 Find the value of k if $(x - 2)$ is a factor of $x^3 - 3x^2 + kx - 10$.

8 Find the remainder when $16x^5 - 20x^4 + 8$ is divided by $(2x - 1)$.

9 $f(x) = 2x^2 + px + q$. Given that $f(-3) = 0$, and $f(4) = 2$:

    a find the value of p and q

    b factorise $f(x)$

10 $h(x) = x^3 + 4x^2 + rx + s$. Given $h(-1) = 0$, and $h(2) = 30$:

    a find the value of r and s

    b find the remainder when $h(x)$ is divided by $(3x - 1)$

11 $g(x) = 2x^3 + 9x^2 - 6x - 5$.

    a Factorise $g(x)$

    b Solve $g(x) = 0$

12 The remainder obtained when $x^3 - 5x^2 + px + 6$ is divided by $(x + 2)$ is equal to the remainder obtained when the same expression is divided by $(x - 3)$.
Find the value of p.

13 The remainder obtained when $x^3 + dx^2 - 5x + 6$ is divided by $(x - 1)$ is twice the remainder obtained when the same expression is divided by $(x + 1)$.
Find the value of d.

14 **a** Show that $(x - 2)$ is a factor of $f(x) = x^3 + x^2 - 5x - 2$.

    b Hence, or otherwise, find the exact solutions of the equation $f(x) = 0$. **E**

15 Given that -1 is a root of the equation $2x^3 - 5x^2 - 4x + 3$, find the two positive roots. **E**

Summary of key points

1 If f(x) is a polynomial and f(a) = 0, then ($x - a$) is a factor of f(x).

2 If f(x) is a polynomial and f$\left(\dfrac{b}{a}\right)$ = 0, then ($ax - b$) is a factor of f(x).

3 If a polynomial f(x) is divided by ($ax - b$) then the remainder is f$\left(\dfrac{b}{a}\right)$.

The sine and cosine rule

In this chapter you will learn how to calculate the lengths of sides, the size of angles, and the area of a triangle.

2.1 The sine rule is:

$$\frac{a}{\sin A} = \frac{b}{\sin B} = \frac{c}{\sin C} \text{ or } \frac{\sin A}{a} = \frac{\sin B}{b} = \frac{\sin C}{c}.$$

You can use the sine rule to find an unknown length when you know two angles and one of the opposite sides.

- When you are finding the length of a side use:

$$\frac{a}{\sin A} = \frac{b}{\sin B} \text{ or } \frac{a}{\sin A} = \frac{c}{\sin C} \text{ or } \frac{b}{\sin B} = \frac{c}{\sin C}$$

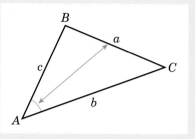

Hint: Note that side a is opposite angle A.

Example 1

In $\triangle ABC$, $AB = 8$ cm, $\angle BAC = 30°$ and $\angle BCA = 40°$. Find BC.

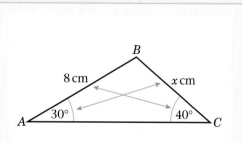

Always draw a diagram and carefully add the data. Here $c = 8$ (cm), $C = 40°$, $A = 30°$, $a = x$ (cm)

In a triangle, the larger a side is, the larger the opposite angle is. Here, as $C > A$, then $c > a$, so you know that $8 > x$.

$$\frac{x}{\sin 30°} = \frac{8}{\sin 40°}$$

Using the sine rule, $\dfrac{a}{\sin A} = \dfrac{c}{\sin C}$.

$$\text{So} \quad x = \frac{8 \sin 30°}{\sin 40°}$$

Multiply throughout by $\sin 30°$.

$$= 6.22$$

Give answer to 3 significant figures.

Example 2

In $\triangle PQR$, $QR = 8.5$ cm, $\angle QPR = 60°$ and $\angle PQR = 20°$. Find PQ.

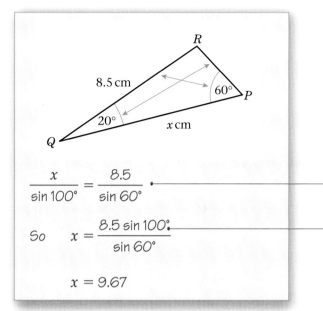

Here $p = 8.5$, $P = 60°$, $Q = 20°$, $r = x$.

To work out PQ you need $\angle R$.

$R = 180° - (60° + 20°) = 100°$. (Angles in a triangle)

As $100° > 60°$, you know that $x > 8.5$.

$$\frac{x}{\sin 100°} = \frac{8.5}{\sin 60°}$$

Using the sine rule, $\dfrac{r}{\sin R} = \dfrac{p}{\sin P}$.

So $\quad x = \dfrac{8.5 \sin 100°}{\sin 60°}$

Multiply throughout by $\sin 100°$.

$$x = 9.67$$

Example 3

Prove the sine rule for a general triangle ABC.

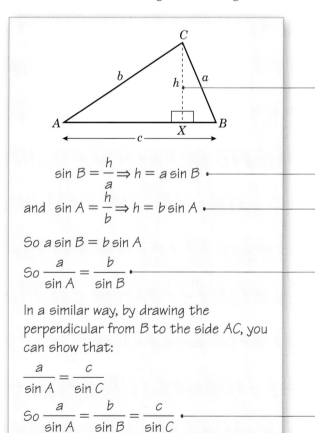

In a general triangle ABC, draw the perpendicular from C to AB.

It meets AB at X.

The length of CX is h.

$$\sin B = \frac{h}{a} \Rightarrow h = a \sin B$$

Use the sine ratio in triangle CBX.

$$\text{and} \quad \sin A = \frac{h}{b} \Rightarrow h = b \sin A$$

Use the sine ratio in triangle CAX.

So $a \sin B = b \sin A$

So $\dfrac{a}{\sin A} = \dfrac{b}{\sin B}$

Divide throughout by $\sin A \sin B$.

In a similar way, by drawing the perpendicular from B to the side AC, you can show that:

$$\frac{a}{\sin A} = \frac{c}{\sin C}$$

So $\dfrac{a}{\sin A} = \dfrac{b}{\sin B} = \dfrac{c}{\sin C}$

This is the sine rule and is true for all triangles.

Exercise 2A   (Give answers to 3 significant figures.)

1 In each of parts **a** to **d**, given values refer to the general triangle:

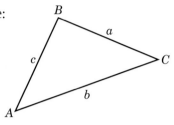

   a Given that $a = 8$ cm, $A = 30°$, $B = 72°$, find b.

   b Given that $a = 24$ cm, $A = 110°$, $C = 22°$, find c.

   c Given that $b = 14.7$ cm, $A = 30°$, $C = 95°$, find a.

   d Given that $c = 9.8$ cm, $B = 68.4°$, $C = 83.7°$, find a.

2 In each of the following triangles calculate the values of x and y.

a 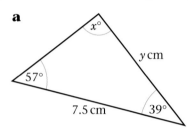   **b**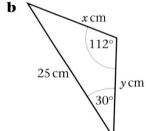

c 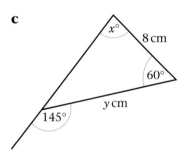   **d**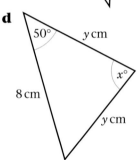

e 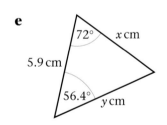   **f**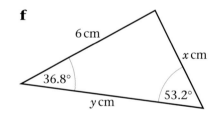

3 In $\triangle PQR$, $QR = \sqrt{3}$ cm, $\angle PQR = 45°$ and $\angle QPR = 60°$. Find **a** PR and **b** PQ.

4 Town B is 6 km, on a bearing of 020°, from town A. Town C is located on a bearing of 055° from town A and on a bearing of 120° from town B. Work out the distance of town C from **a** town A and **b** town B.

5 In the diagram $AD = DB = 5$ cm, $\angle ABC = 43°$ and $\angle ACB = 72°$.
Calculate **a** AB and **b** CD.

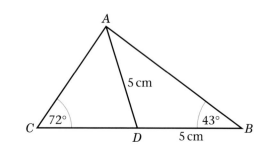

2.2 You can use the sine rule to find an unknown angle in a triangle when you know two sides and one of their opposite angles.

When you are finding an angle use:

$$\frac{\sin A}{a} = \frac{\sin B}{b} \quad \text{or} \quad \frac{\sin A}{a} = \frac{\sin C}{c} \quad \text{or} \quad \frac{\sin B}{b} = \frac{\sin C}{c}$$

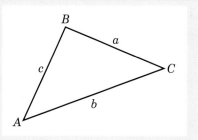

Example 4

In $\triangle ABC$, $AB = 4$ cm, $AC = 12$ cm and $\angle ABC = 64°$. Find $\angle ACB$.

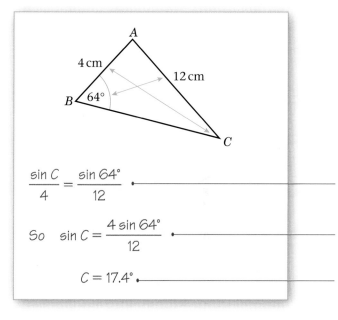

Here $b = 12$ cm, $c = 4$ cm, $B = 64°$.

$$\frac{\sin C}{4} = \frac{\sin 64°}{12}$$

As you need to find angle C, use the sine rule $\frac{\sin C}{c} = \frac{\sin B}{b}$.

So $\quad \sin C = \dfrac{4 \sin 64°}{12}$

As $4 < 12$, you know that $C < 64°$.

$$C = 17.4°$$

$C = \sin^{-1}\left(\dfrac{4 \sin 64°}{12}\right)$.

Example 5

In $\triangle ABC$, $AB = 3.8$ cm, $BC = 5.2$ cm and $\angle BAC = 35°$. Find $\angle ABC$.

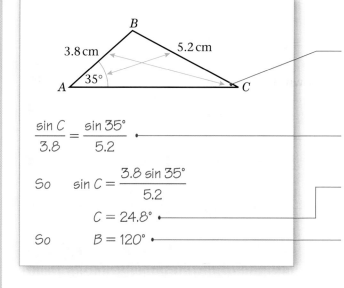

Here $a = 5.2$ cm, $c = 3.8$ cm and $A = 35°$.
You first need to find angle C.

$$\frac{\sin C}{3.8} = \frac{\sin 35°}{5.2}$$

Use $\dfrac{\sin C}{c} = \dfrac{\sin A}{a}$.

So $\quad \sin C = \dfrac{3.8 \sin 35°}{5.2}$

You know that $C < 35°$.

$$C = 24.8°$$

So $\quad B = 120°$

$B = 180° - (24.8° + 35°) = 120.2°$, which rounds to $120°$ to 3 significant figures.

Exercise 2B

(*Note:* Give answers to 3 significant figures, unless they are exact.)

1 In each of the following sets of data for a triangle *ABC*, find the value of *x*:

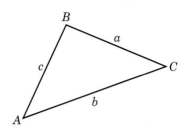

   a *AB* = 6 cm, *BC* = 9 cm, ∠*BAC* = 117°, ∠*ACB* = *x*°.

   b *AC* = 11 cm, *BC* = 10 cm, ∠*ABC* = 40°, ∠*CAB* = *x*°.

   c *AB* = 6 cm, *BC* = 8 cm, ∠*BAC* = 60°, ∠*ACB* = *x*°.

   d *AB* = 8.7 cm, *AC* = 10.8 cm, ∠*ABC* = 28°, ∠*BAC* = *x*°.

2 In each of the diagrams shown below, work out the value of *x*:

a
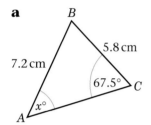

b
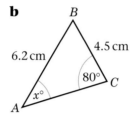

c
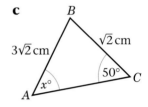

d
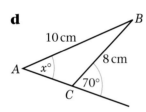

e
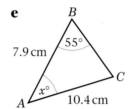

f
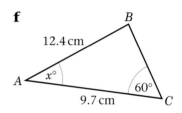

3 In △*PQR*, *PQ* = 15 cm, *QR* = 12 cm and ∠*PRQ* = 75°. Find the two remaining angles.

4 In each of the following diagrams work out the values of *x* and *y*:

a
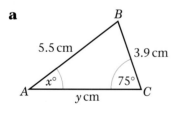

b
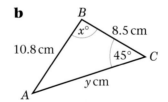

c
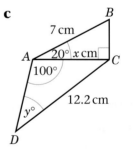

d
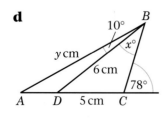

e
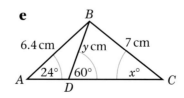

f
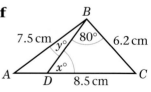

5 In △*ABC*, *AB* = *x* cm, *BC* = (4 − *x*) cm, ∠*BAC* = *y*° and ∠*BCA* = 30°.

   Given that $\sin y° = \dfrac{1}{\sqrt{2}}$, show that $x = 4(\sqrt{2} - 1)$.

2.3 You can sometimes find two solutions for a missing angle.

- When the angle you are finding is larger than the given angle, there are two possible results. This is because you can draw two possible triangles with the data.
- In general, $\sin(180 - x)° = \sin x°$. For example $\sin 30° = \sin 150°$.

Example 6

In $\triangle ABC$, $AB = 4$ cm, $AC = 3$ cm and $\angle ABC = 44°$. Work out the two possible values of $\angle ACB$.

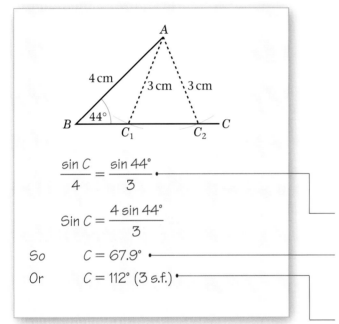

$$\frac{\sin C}{4} = \frac{\sin 44°}{3}$$

$$\sin C = \frac{4 \sin 44°}{3}$$

So $\quad C = 67.9°$

Or $\quad C = 112°$ (3 s.f.)

Here $\angle ACB > \angle ABC$, as $AB > AC$, and so there will be two possible results. The diagram shows why:

With $\angle ABC = 44°$ and $AB = 4$ cm drawn, imagine putting a pair of compasses at A, then drawing an arc with centre A and radius 3 cm. This will intersect BC at C_1 and C_2 showing that there are two triangles ABC_1 and ABC_2 where $b = 3$ cm, $c = 4$ cm and $B = 44°$.

(This would not happen if $AC > 4$ cm.)

Use $\dfrac{\sin C}{c} = \dfrac{\sin B}{b}$, where $b = 3$, $c = 4$, $B = 44°$.

This is the value your calculator will give to 3 s.f., which corresponds to $\triangle ABC_2$.

As $\sin(180 - x)° = \sin x°$, $C = 180 - 67.9° = 112.1°$ is another possible answer. This corresponds to $\triangle ABC_1$.

Exercise 2C

(Give answers to 3 significant figures.)

1 In $\triangle ABC$, $BC = 6$ cm, $AC = 4.5$ cm and $\angle ABC = 45°$:

 a Calculate the two possible values of $\angle BAC$.

 b Draw a diagram to illustrate your answers.

2 In each of the diagrams shown below, calculate the possible values of x and the corresponding values of y:

a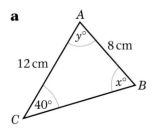

b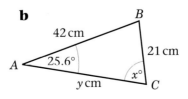

c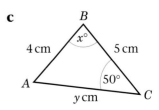

3 In each of the following cases △*ABC* has ∠*ABC* = 30° and *AB* = 10 cm:

    a Calculate the least possible length that *AC* could be.

    b Given that *AC* = 12 cm, calculate ∠*ACB*.

    c Given instead that *AC* = 7 cm, calculate the two possible values of ∠*ACB*.

4 Triangle *ABC* is such that *AB* = 4 cm, *BC* = 6 cm and ∠*ACB* = 36°. Show that one of the possible values of ∠*ABC* is 25.8° (to 3 s.f.). Using this value, calculate the length of *AC*.

5 Two triangles *ABC* are such that *AB* = 4.5 cm, *BC* = 6.8 cm and ∠*ACB* = 30°. Work out the value of the largest angle in each of the triangles.

2.4 **The cosine rule is:**

$$a^2 = b^2 + c^2 - 2bc \cos A \text{ or } \cos A = \frac{b^2 + c^2 - a^2}{2bc}$$

You can use the cosine rule to find an unknown side in a triangle when you know the lengths of two sides and the size of the angle between the sides.

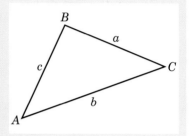

- **In a question use** $a^2 = b^2 + c^2 - 2bc \cos A$ **to find** *a*, **given** *b*, *c* **and** *A*.
  or $b^2 = a^2 + c^2 - 2ac \cos B$ **to find** *b*, **given** *a*, *c* **and** *B*.
  or $c^2 = a^2 + b^2 - 2ab \cos C$ **to find** *c*, **given** *a*, *b* **and** *C*.

Example   7

Calculate the length of the side *AB* of the triangle *ABC* in which *AC* = 6.5 cm, *BC* = 8.7 cm and ∠*ACB* = 100°.

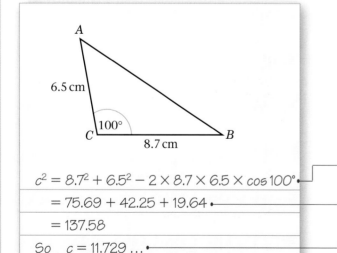

You have been given *a*, *b* and angle *C*, so use the cosine rule:
$$c^2 = a^2 + b^2 - 2ab \cos C$$
to find *c*.

$c^2 = 8.7^2 + 6.5^2 - 2 \times 8.7 \times 6.5 \times \cos 100°$ Carefully set out this line of working before using your calculator.

     $= 75.69 + 42.25 + 19.64$ This line may be omitted.

     $= 137.58$

So $c = 11.729\ldots$ Find the square root.

So $AB = 11.7$ cm (3 s.f.)

Example 8

Coastguard station B is 8 km, on a bearing of 060°, from coastguard station A. A ship C is 4.8 km, on a bearing of 018°, away from A.
Calculate how far C is from B.

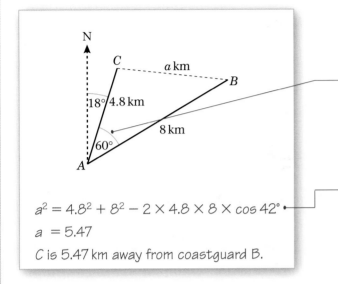

Carefully transfer the given data to a diagram.
In $\triangle ABC$, $\angle CAB = 60° - 18° = 42°$.

You now have $b = 4.8$ km, $c = 8$ km and $A = 42°$.
Use the cosine rule $a^2 = b^2 + c^2 - 2bc \cos A$.

$a^2 = 4.8^2 + 8^2 - 2 \times 4.8 \times 8 \times \cos 42°$

$a = 5.47$

C is 5.47 km away from coastguard B.

Example 9

In $\triangle ABC$, $AB = x$ cm, $BC = (x + 2)$ cm, $AC = 5$ cm and $\angle ABC = 60°$.
Find the value of x.

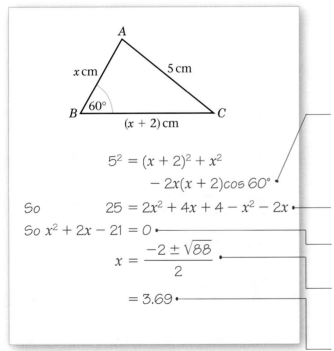

The given data here are
$a = (x + 2)$, $c = x$, $b = 5$, $B = 60°$.

The sine rule cannot be used, but you can use $b^2 = a^2 + c^2 - 2ac \cos B$.

$5^2 = (x + 2)^2 + x^2$
$\qquad - 2x(x + 2)\cos 60°$

So $\qquad 25 = 2x^2 + 4x + 4 - x^2 - 2x$

$(x + 2)^2 = x^2 + 4x + 4$; $\cos 60° = \frac{1}{2}$.

So $x^2 + 2x - 21 = 0$

Rearrange to the form $ax^2 + bx + c = 0$.

$$x = \frac{-2 \pm \sqrt{88}}{2}$$

Use the quadratic equation formula, where $b^2 - 4ac = 2^2 - 4(1)(-21) = 4 + 84 = 88$.

$= 3.69$

As $AB = x$ cm, x cannot be negative.

Example 10

Prove the cosine rule for a general triangle *ABC*.

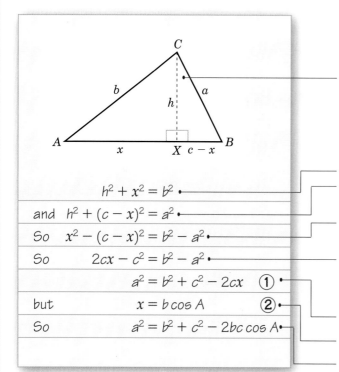

The perpendicular from *C* to side *AB* is drawn and it meets *AB* at *X*.

The length of *CX* is *h*.

The length of *AX* is *x*, so *BX* = *c* − *x*.

$$h^2 + x^2 = b^2$$

and $h^2 + (c - x)^2 = a^2$

So $x^2 - (c - x)^2 = b^2 - a^2$

So $2cx - c^2 = b^2 - a^2$

$a^2 = b^2 + c^2 - 2cx$ ①

but $x = b \cos A$ ②

So $a^2 = b^2 + c^2 - 2bc \cos A$

Use Pythagoras' theorem in △*CAX*.
Use Pythagoras' theorem in △*CBX*.

Subtract the two equations.

$(c - x)^2 = c^2 - 2cx + x^2$.
So $x^2 - (c - x)^2 = x^2 - c^2 + 2cx - x^2$.

Rearrange.

Use the cosine ratio $\cos A = \dfrac{x}{b}$ in △*CAX*.

Combine ① and ②. This is the cosine rule.

Exercise 2D

(*Note:* Give answers to 3 significant figures, where appropriate.)

1 In each of the following triangles calculate the length of the third side:

a
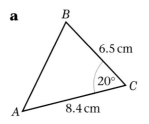

b
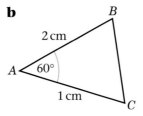

c
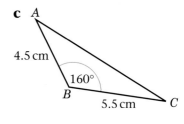

d
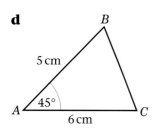

e
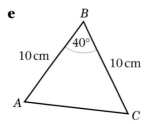

f
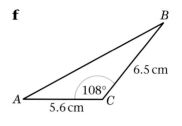

2 From a point *A* a boat sails due north for 7 km to *B*. The boat leaves *B* and moves on a bearing of 100° for 10 km until it reaches *C*. Calculate the distance of *C* from *A*.

3 The distance from the tee, T, to the flag, F, on a particular hole on a golf course is 494 yards. A golfer's tee shot travels 220 yards and lands at the point S, where $\angle STF = 22°$. Calculate how far the ball is from the flag.

4 In $\triangle ABC$, $AB = (x - 3)$ cm, $BC = (x + 3)$ cm, $AC = 8$ cm and $\angle BAC = 60°$. Use the cosine rule to find the value of x.

5 In $\triangle ABC$, $AB = x$ cm, $BC = (x - 4)$ cm, $AC = 10$ cm and $\angle BAC = 60°$. Calculate the value of x.

6 In $\triangle ABC$, $AB = (5 - x)$ cm, $BC = (4 + x)$ cm, $\angle ABC = 120°$ and $AC = y$ cm.

  a Show that $y^2 = x^2 - x + 61$.

  b Use the method of completing the square to find the minimum value of y^2, and give the value of x for which this occurs.

> **Hint for question 6b:**
> Completing the square is in Book C1, Chapter 2.

2.5 You can use the cosine rule to find an unknown angle in a triangle if you know the lengths of all three sides.

■ You can find an unknown angle using a rearranged form of the cosine rule:

$$\cos A = \frac{b^2 + c^2 - a^2}{2bc}$$

or $$\cos B = \frac{a^2 + c^2 - b^2}{2ac}$$

or $$\cos C = \frac{a^2 + b^2 - c^2}{2ab}$$

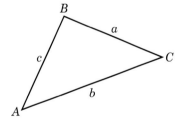

Example 11

Rearrange the equation $a^2 = b^2 + c^2 - 2bc \cos A$ in the form $\cos A = \ldots$

$$a^2 = b^2 + c^2 - 2bc \cos A$$

So $2bc \cos A = b^2 + c^2 - a^2$

So $\cos A = \dfrac{b^2 + c^2 - a^2}{2bc}$ —— Divide throughout by $2bc$.

Example 12

In $\triangle PQR$, $PQ = 5.9$ cm, $QR = 8.2$ cm and $PR = 10.6$ cm.
Calculate the size of $\angle PQR$.

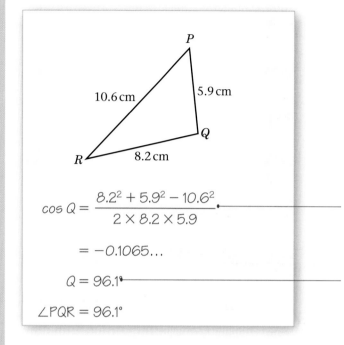

Here $p = 8.2$ cm, $r = 5.9$ cm, $q = 10.6$ cm, and you have to find angle Q.

Use the cosine rule $\cos Q = \dfrac{p^2 + r^2 - q^2}{2pr}$

$$\cos Q = \frac{8.2^2 + 5.9^2 - 10.6^2}{2 \times 8.2 \times 5.9}$$

$$= -0.1065\ldots$$

$$Q = 96.1°$$

$Q = \cos^{-1}(-0.1065\ldots)$

$$\angle PQR = 96.1°$$

Example 13

Find the size of the smallest angle in a triangle whose sides have lengths 3 cm, 5 cm and 6 cm.

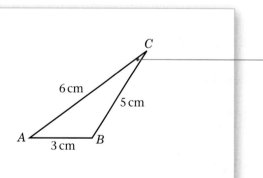

Label the triangle ABC.
The smallest angle is opposite the smallest side so angle C is required.

$$\cos C = \frac{6^2 + 5^2 - 3^2}{2 \times 6 \times 5}$$

Use the cosine rule $\cos C = \dfrac{a^2 + b^2 - c^2}{2ab}$

$$C = 29.9°$$

The size of the smallest angle is 29.9°.

Exercise 2E

(Give answers to 3 significant figures.)

1 In the following triangles calculate the size of the angle marked *:

a
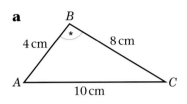

b
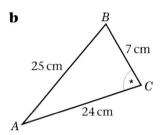

c
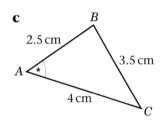

d
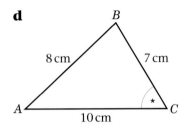

e
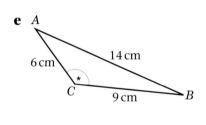

f
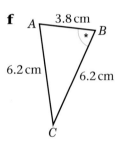

2 A helicopter flies on a bearing of 080° from A to B, where $AB = 50$ km.
It then flies for 60 km to a point C.
Given that C is 80 km from A, calculate the bearing of C from A.

3 In $\triangle ABC$, $AB = 5$ cm, $BC = 6$ cm and $AC = 10$ cm.
Calculate the value of the smallest angle.

4 In $\triangle ABC$, $AB = 9.3$ cm, $BC = 6.2$ cm and $AC = 12.7$ cm.
Calculate the value of the largest angle.

5 The lengths of the sides of a triangle are in the ratio $2 : 3 : 4$.
Calculate the value of the largest angle.

6 In $\triangle ABC$, $AB = x$ cm, $BC = 5$ cm and $AC = (10 - x)$ cm:

a Show that $\cos \angle ABC = \dfrac{4x - 15}{2x}$.

b Given that $\cos \angle ABC = -\frac{1}{7}$, work out the value of x.

2.6 **You need to be able to use the sine rule, the cosine rule, the trigonometric ratios sin, cos and tan, and Pythagoras' theorem to solve problems.**

In triangle work involving trigonometric calculations, the following strategy might help you.

- When the triangle is right-angled or isosceles it is better to use sine, cosine, tangent or Pythagoras' theorem.

$\sin A = \dfrac{a}{b}$

$\cos A = \dfrac{c}{b}$

$\tan A = \dfrac{a}{c}$

$a^2 + c^2 = b^2$

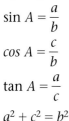 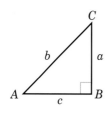

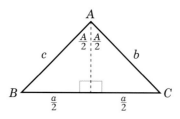

The line of symmetry produces two right-angled triangles.

- Use the **cosine rule** when you are given either **two sides and the angle between them** or **three sides**.

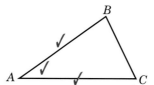 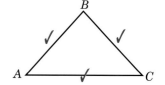

- For other combinations of given data, use the **sine rule**.
- When you have used the cosine rule once, it is generally better not to use it again, as the cosine rule involves more calculations and so may introduce more rounding errors.

Example 14

In $\triangle ABC$, $AB = 5.2$ cm, $BC = 6.4$ cm and $AC = 3.6$ cm.
Calculate the angles of the triangle.

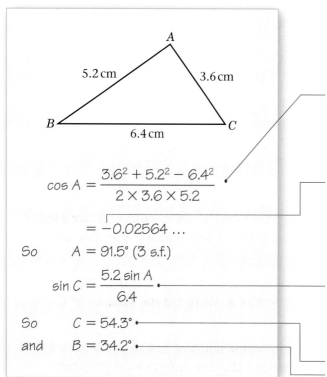

It is always better to work out the largest angle first, if you have a choice.

Here that is angle A, so use the cosine rule $\cos A = \dfrac{b^2 + c^2 - a^2}{2bc}$.

Negative sign indicates an obtuse angle: $\angle BAC = 91.5°$

Store the calculator value for A.

To find a second angle it is better to use the sine rule rather than a second cosine rule.

Use $\dfrac{\sin C}{c} = \dfrac{\sin A}{a}$, and use the stored value for A.

$\angle ACB = 54.3°$.
$\angle ABC = 180° - (91.5 + 54.3)°$.

Exercise 2F

(*Note:* Try to use the neatest method, and give answers to 3 significant figures.)

1 In each triangle below find the values of *x*, *y* and *z*.

a
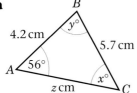

b
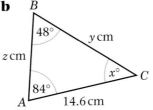

c
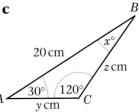

d
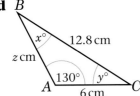

e
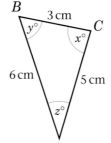

f
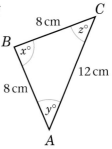

g
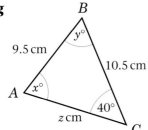

h
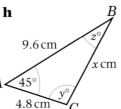

i
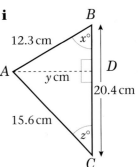

2 Calculate the size of the remaining angles and the length of the third side in the following triangles:

a
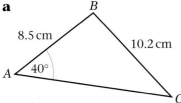

b
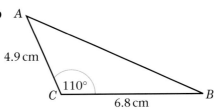

3 A hiker walks due north from *A* and after 8 km reaches *B*. She then walks a further 8 km on a bearing of 120° to *C*. Work out **a** the distance from *A* to *C* and **b** the bearing of *C* from *A*.

4 A helicopter flies on a bearing of 200° from *A* to *B*, where *AB* = 70 km. It then flies on a bearing of 150° from *B* to *C*, where *C* is due south of *A*. Work out the distance of *C* from *A*.

5 Two radar stations *A* and *B* are 16 km apart and *A* is due north of *B*. A ship is known to be on a bearing of 150° from *A* and 10 km from *B*. Show that this information gives two positions for the ship, and calculate the distance between these two positions.

6 Find x in each of the following diagrams:

a

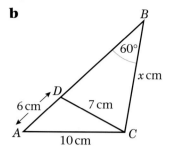

b

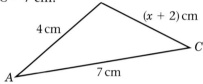

c

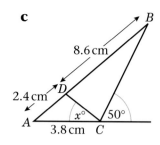

7 In $\triangle ABC$, shown right, $AB = 4$ cm, $BC = (x + 2)$ cm and $AC = 7$ cm.

   a Explain how you know that $1 < x < 9$.

   b Work out the value of x for the cases when

      i $\angle ABC = 60°$ and

      ii $\angle ABC = 45°$, giving your answers to 3 significant figures.

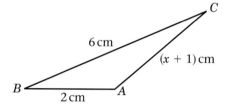

8 In the triangle shown right, $\cos \angle ABC = \frac{5}{8}$.
Calculate the value of x.

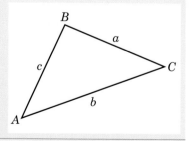

9 In $\triangle ABC$, $AB = \sqrt{2}$ cm, $BC = \sqrt{3}$ cm and $\angle BAC = 60°$. Show that $\angle ACB = 45°$ and find AC.

10 In $\triangle ABC$, $AB = (2 - x)$ cm, $BC = (x + 1)$ cm and $\angle ABC = 120°$:

   a Show that $AC^2 = x^2 - x + 7$.

   b Find the value of x for which AC has a minimum value.

11 Triangle ABC is such that $BC = 5\sqrt{2}$ cm, $\angle ABC = 30°$ and $\angle BAC = \theta$, where $\sin \theta = \dfrac{\sqrt{5}}{8}$.

   Work out the length of AC, giving your answer in the form $a\sqrt{b}$, where a and b are integers.

12 The perimeter of $\triangle ABC = 15$ cm. Given that $AB = 7$ cm and $\angle BAC = 60°$, find the lengths of AC and BC.

2.7 **You can calculate the area of a triangle using the formula:**

Area of a triangle $= \frac{1}{2}ab \sin C$ or $\frac{1}{2}ac \sin B$ or $\frac{1}{2}bc \sin A$.

You can use this formula when you know the lengths of two sides and the size of the angle between them.

Example 15

Show that the area of this triangle is $\frac{1}{2}ab \sin C$.

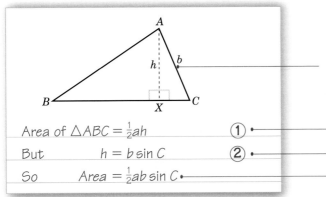

Area of $\triangle ABC = \frac{1}{2}ah$ ① • ——— The perpendicular from A to BC is drawn and it meets BC at X. The length of $AX = h$.

But $h = b \sin C$ ② • ——— Area of triangle $= \frac{1}{2} \times$ base $\times$ height.

So   Area $= \frac{1}{2}ab \sin C$ • ——— Use the sine ratio $\sin C = \dfrac{h}{b}$ in $\triangle AXC$.

Substitute ② into ①.

Example 16

Work out the area of the triangle shown below.

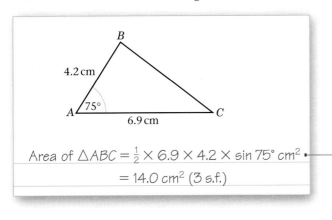

Area of $\triangle ABC = \frac{1}{2} \times 6.9 \times 4.2 \times \sin 75°$ cm² • ——— Here $b = 6.9$ cm, $c = 4.2$ cm and angle $A = 75°$, so use:

$= 14.0$ cm² (3 s.f.)

Area $= \frac{1}{2}bc \sin A$.

Example 17

In $\triangle ABC$, $AB = 5$ cm, $BC = 6$ cm and $\angle ABC = x°$. Given that the area of $\triangle ABC$ is 12 cm² and that AC is the longest side, find the value of x.

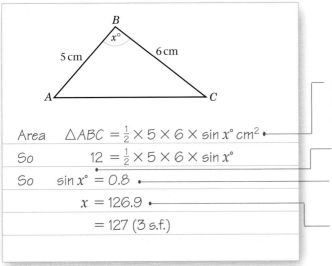

Area $\triangle ABC = \frac{1}{2} \times 5 \times 6 \times \sin x°$ cm² • ——— Here $a = 6$ cm, $c = 5$ cm and angle $B = x°$, so use:

Area $= \frac{1}{2}ac \sin B$.

So $12 = \frac{1}{2} \times 5 \times 6 \times \sin x°$ • ——— Area of $\triangle ABC$ is 12 cm².

So $\sin x° = 0.8$ • ——— Sin $x° = \frac{12}{15}$.

$x = 126.9$ •

$= 127$ (3 s.f.)

There are two values of x for which $\sin x° = 0.8$, 53.1 and 126.9, but here you know B is the largest angle because AC is the largest side.

Exercise 2G

1 Calculate the area of the following triangles:

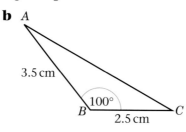

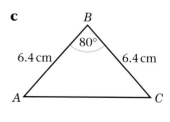

a B, A, $8.6\,\text{cm}$, C, $45°$, $7.8\,\text{cm}$, A

b A, $3.5\,\text{cm}$, $100°$, B, $2.5\,\text{cm}$, C

c B, $80°$, $6.4\,\text{cm}$, $6.4\,\text{cm}$, A, C

2 Work out the possible values of x in the following triangles:

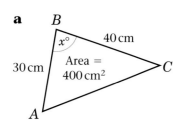

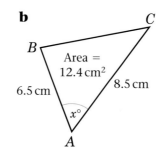

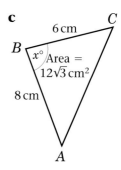

a B, $x°$, $40\,\text{cm}$, $30\,\text{cm}$, Area = $400\,\text{cm}^2$, C, A

b C, B, Area = $12.4\,\text{cm}^2$, $6.5\,\text{cm}$, $8.5\,\text{cm}$, $x°$, A

c C, $6\,\text{cm}$, B, $x°$, Area = $12\sqrt{3}\,\text{cm}^2$, $8\,\text{cm}$, A

3 A fenced triangular plot of ground has area $1200\,\text{m}^2$. The fences along the two smaller sides are $60\,\text{m}$ and $80\,\text{m}$ respectively and the angle between them is $\theta°$. Show that $\theta = 150$, and work out the total length of fencing.

4 In triangle ABC, shown right,
$BC = (x + 2)$ cm, $AC = x$ cm and
$\angle BCA = 150°$.
Given that the area of the triangle
is $5\,\text{cm}^2$, work out the value of x,
giving your answer to 3 significant figures.

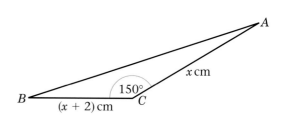

5 In $\triangle PQR$, $PQ = (x + 2)$ cm, $PR = (5 - x)$ cm and $\angle QPR = 30°$.
The area of the triangle is $A\ \text{cm}^2$:

a Show that $A = \frac{1}{4}(10 + 3x - x^2)$.

b Use the method of completing the square, or otherwise, to find the maximum value of A, and give the corresponding value of x.

6 In $\triangle ABC$, $AB = x$ cm, $AC = (5 + x)$ cm and $\angle BAC = 150°$. Given that the area of the triangle is $3\frac{3}{4}\,\text{cm}^2$:

a Show that x satisfies the equation $x^2 + 5x - 15 = 0$.

b Calculate the value of x, giving your answer to 3 significant figures.

Mixed exercise 2H

(Give non-exact answers to 3 significant figures.)

1 The area of a triangle is 10 cm². The angle between two of the sides, of length 6 cm and 8 cm respectively, is obtuse. Work out:

a The size of this angle.

b The length of the third side.

2 In each triangle below, find the value of x and the area of the triangle:

a
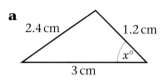

b
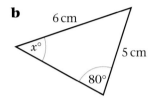

c
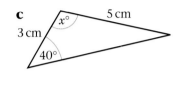

3 The sides of a triangle are 3 cm, 5 cm and 7 cm respectively. Show that the largest angle is 120°, and find the area of the triangle.

4 In each of the figures below calculate the total area:

a
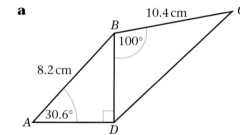

b
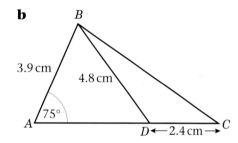

5 In $\triangle ABC$, $AB = 10$ cm, $BC = a\sqrt{3}$ cm, $AC = 5\sqrt{13}$ cm and $\angle ABC = 150°$. Calculate:

a The value of a.

b The exact area of $\triangle ABC$.

6 In a triangle, the largest side has length 2 cm and one of the other sides has length $\sqrt{2}$ cm. Given that the area of the triangle is 1 cm², show that the triangle is right-angled and isosceles.

7 The three points A, B and C, with coordinates $A(0, 1)$, $B(3, 4)$ and $C(1, 3)$ respectively, are joined to form a triangle:

a Show that $\cos \angle ACB = -\frac{4}{5}$.

b Calculate the area of $\triangle ABC$.

8 The longest side of a triangle has length $(2x - 1)$ cm. The other sides have lengths $(x - 1)$ cm and $(x + 1)$ cm. Given that the largest angle is 120°, work out:

a the value of x and **b** the area of the triangle.

Summary of key points

1 The sine rule is

$$\frac{a}{\sin A} = \frac{b}{\sin B} = \frac{c}{\sin C} \quad or \quad \frac{\sin A}{a} = \frac{\sin B}{b} = \frac{\sin C}{c}$$

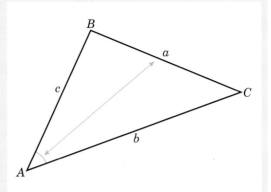

2 You can use the sine rule to find an unknown side in a triangle if you know two angles and the length of one of their opposite sides.

3 You can use the sine rule to find an unknown angle in a triangle if you know the lengths of two sides and one of their opposite angles.

4 The cosine rule is

$$a^2 = b^2 + c^2 - 2bc \cos A \quad or \quad b^2 = a^2 + c^2 - 2ac \cos B \quad or \quad c^2 = a^2 + b^2 - 2ab \cos C$$

5 You can use the cosine rule to find an unknown side in a triangle if you know the lengths of two sides and the angle between them.

6 You can use the cosine rule to find an unknown angle if you know the lengths of all three sides.

7 You can find an unknown angle using a rearranged form of the cosine rule:

$$\cos A = \frac{b^2 + c^2 - a^2}{2bc} \quad or \quad \cos B = \frac{a^2 + c^2 - b^2}{2ac} \quad or \quad \cos C = \frac{a^2 + b^2 - c^2}{2ab}$$

8 You can find the area of a triangle using the formula

$$\text{area} = \tfrac{1}{2}ab \sin C$$

if you know the length of two sides (a and b) and the value of the angle C between them.

3 Exponentials and logarithms

This chapter introduces you to exponential functions and logarithms. You will learn how to sketch the graph of an exponential function, how to write an expression as a logarithm and how to use logarithms to solve equations.

3.1 You need to be familiar with the function $y = a^x$ ($a > 0$) and to know the shape of its graph.

As an example, look at a table of values for $y = 2^x$:

x	-3	-2	-1	0	1	2	3
y	$\frac{1}{8}$	$\frac{1}{4}$	$\frac{1}{2}$	1	2	4	8

Hint: In an expression such as 2^x, the x can be called a power, or an index, or an exponent. You will see in Section 3.2 that it can also be thought of as a logarithm. A function that involves variable power such as x is called an exponential function.

Note that

$$2^0 = 1 \text{ (in fact } a^0 = 1 \text{ always if } a > 0)$$

and $2^{-3} = \dfrac{1}{2^3} = \dfrac{1}{8}$ (a negative index implies the 'reciprocal' of a positive index)

Hint: See Book C1, Chapter 1 for the rules of indices.

The graph of $y = 2^x$ looks like this:

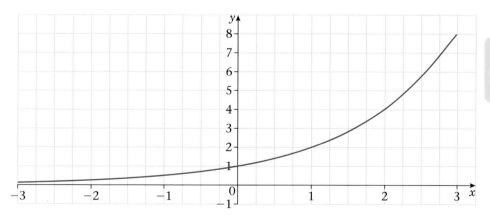

Hint: The x-axis is an asymptote to the curve.

Other graphs of the type $y = a^x$ are of a similar shape, always passing through $(0, 1)$.

Example 1

a On the same axes sketch the graphs of $y = 3^x$, $y = 2^x$ and $y = 1.5^x$.

b On another set of axes sketch the graphs of $y = (\frac{1}{2})^x$ and $y = 2^x$.

a For all three graphs, $y = 1$ when $x = 0$. •————————————— $a^0 = 1$

When $x > 0$, $3^x > 2^x > 1.5^x$.•———

When $x < 0$, $3^x < 2^x < 1.5^x$.•———

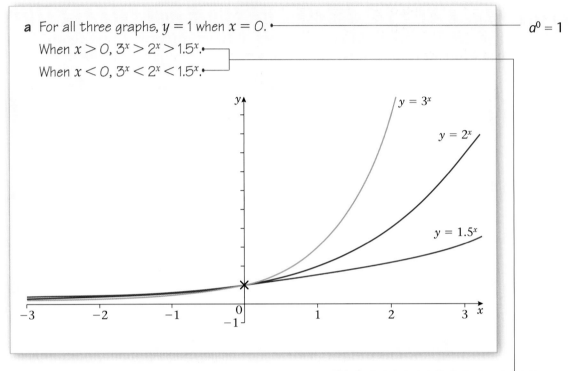

Work out the relative positions of the three graphs.

b $\dfrac{1}{2} = 2^{-1}$

So $y = \left(\dfrac{1}{2}\right)^x$ is the same as $y = (2^{-1})^x = 2^{-x}$. •————————————— $(a^m)^n = a^{mn}$

So the graph of $y = \left(\dfrac{1}{2}\right)^x$ is a reflection in the y-axis of

the graph of $y = 2^x$.

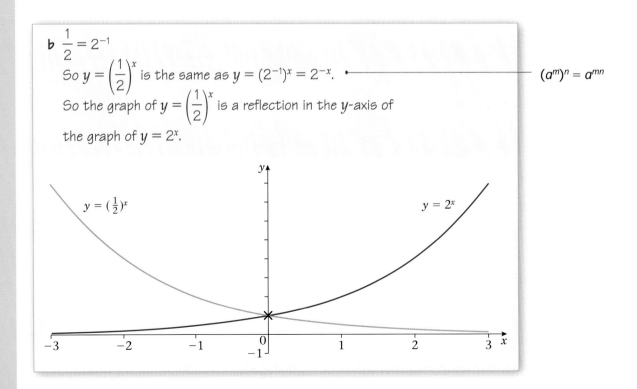

Exercise 3A

1 **a** Draw an accurate graph of $y = (1.7)^x$, for $-4 \leqslant x \leqslant 4$.
   b Use your graph to solve the equation $(1.7)^x = 4$.

2 **a** Draw an accurate graph of $y = (0.6)^x$, for $-4 \leqslant x \leqslant 4$.
   b Use your graph to solve the equation $(0.6)^x = 2$

3 Sketch the graph of $y = 1^x$.

3.2 You need to know how to write an expression as a logarithm.

■ $\log_a n = x$ means that $a^x = n$, where a is called the base of the logarithm.

Example 2

Write as a logarithm $2^5 = 32$.

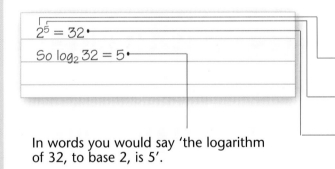

Here $a = 2$, $x = 5$, $n = 32$.

Base.

Logarithm.

$2^5 = 32$

So $\log_2 32 = 5$

In words you would say 'the logarithm of 32, to base 2, is 5'.

In words, you would say '2 to the power 5 equals 32'.

Example 3

Rewrite as a logarithm:
a $10^3 = 1000$
b $5^4 = 625$
c $2^{10} = 1024$

a $\log_{10} 1000 = 3$
b $\log_5 625 = 4$
c $\log_2 1024 = 10$

■ $\log_a 1 = 0$ $(a > 0)$ ———————— Because $a^0 = 1$.

■ $\log_a a = 1$ $(a > 0)$ ———————— Because $a^1 = a$.

Example 4

Find the value of

a $\log_3 81$ **b** $\log_4 0.25$ **c** $\log_{0.5} 4$ **d** $\log_a (a^5)$

a $\log_3 81 = 4$ — Because $3^4 = 81$.

b $\log_4 0.25 = -1$ — Because $4^{-1} = \frac{1}{4} = 0.25$.

c $\log_{0.5} 4 = -2$ — Because $0.5^{-2} = \left(\frac{1}{2}\right)^{-2} = 2^2 = 4$.

d $\log_a (a^5) = 5$ — Because $a^5 = a^5$!

Exercise 3B

1 Rewrite as a logarithm:

a $4^4 = 256$ **b** $3^{-2} = \frac{1}{9}$ **c** $10^6 = 1\,000\,000$

d $11^1 = 11$ **e** $(0.2)^3 = 0.008$

2 Rewrite using a power:

a $\log_2 16 = 4$ **b** $\log_5 25 = 2$ **c** $\log_9 3 = \frac{1}{2}$

d $\log_5 0.2 = -1$ **e** $\log_{10} 100\,000 = 5$

3 Find the value of:

a $\log_2 8$ **b** $\log_5 25$

c $\log_{10} 10\,000\,000$ **d** $\log_{12} 12$

e $\log_3 729$ **f** $\log_{10} \sqrt{10}$

g $\log_4 (0.25)$ **h** $\log_{0.25} 16$

i $\log_a (a^{10})$ **j** $\log_{\left(\frac{2}{3}\right)}\left(\frac{9}{4}\right)$

4 Find the value of x for which:

a $\log_5 x = 4$ **b** $\log_x 81 = 2$

c $\log_7 x = 1$ **d** $\log_x (2x) = 2$

3.3 You need to be able to calculate logarithms to the base 10 using your calculator.

Example 5

Find the value of x for which $10^x = 500$.

$10^x = 500$

So $\log_{10} 500 = x$ — Since $10^2 = 100$ and $10^3 = 1000$, x must be somewhere between 2 and 3.

$x = \log_{10} 500$

$= 2.70$ (to 3 s.f.) — The log (or lg) button on your calculator gives values of logs to base 10.

Exercise 3C

Find from your calculator the value to 3 s.f. of:

1 $\log_{10} 20$ **2** $\log_{10} 4$

3 $\log_{10} 7000$ **4** $\log_{10} 0.786$

5 $\log_{10} 11$ **6** $\log_{10} 35.3$

7 $\log_{10} 0.3$ **8** $\log_{10} 999$

3.4 You need to know the laws of logarithms.

Suppose that	$\log_a x = b$ and $\log_a y = c$
Rewriting with powers:	$a^b = x$ and $a^c = y$
Multiplying:	$xy = a^b \times a^c = a^{b+c}$ (see Book C1, Chapter 1)
	$xy = a^{b+c}$
Rewriting as a logarithm:	$\log_a xy = b + c$

■ $\log_a xy = \log_a x + \log_a y$ (the multiplication law)

It can also be shown that:

■ $\log_a \left(\dfrac{x}{y}\right) = \log_a x - \log_a y$ (the division law)   Remember: $\dfrac{a^b}{a^c} = a^b \div a^c = a^{b-c}$

■ $\log_a (x)^k = k \log_a x$ (the power law)

                                        Remember: $(a^b)^k = a^{bk}$

Note: You need to learn and remember the above three laws of logarithms.

Since $\dfrac{1}{x} = x^{-1}$, the power rule shows that $\log_a \left(\dfrac{1}{x}\right) = \log_a (x^{-1}) = -\log_a x$.

■ $\log_a \left(\dfrac{1}{x}\right) = -\log_a x$

Example 6

Write as a single logarithm:

a $\log_3 6 + \log_3 7$ **b** $\log_2 15 - \log_2 3$

c $2 \log_5 3 + 3 \log_5 2$ **d** $\log_{10} 3 - 4 \log_{10} \left(\tfrac{1}{2}\right)$

---

a $\log_3 (6 \times 7)$

     $= \log_3 42$ ←——————————— Use the multiplication law.

---

b $\log_2 (15 \div 3)$

     $= \log_2 5$ ←——————————— Use the division law.

c $\quad 2\log_5 3 = \log_5 (3^2) = \log_5 9$

$\quad 3\log_5 2 = \log_5 (2^3) = \log_5 8$

$\quad \log_5 9 + \log_5 8 = \log_5 72$

First apply the power law to both parts of the expression.
Then use the multiplication law.

d $\quad 4\log_{10}\left(\dfrac{1}{2}\right) = \log_{10}\left(\dfrac{1}{2}\right)^4 = \log_{10}\left(\dfrac{1}{16}\right)$

$\quad \log_{10} 3 - \log_{10}\left(\dfrac{1}{16}\right) = \log_{10}\left(3 \div \dfrac{1}{16}\right)$

$\quad\quad\quad\quad\quad = \log_{10} 48$

Use the power law first.
Then use the division law.

Example 7

Write in terms of $\log_a x$, $\log_a y$ and $\log_a z$

a $\log_a (x^2 y z^3)$ **b** $\log_a\left(\dfrac{x}{y^3}\right)$ **c** $\log_a\left(\dfrac{x\sqrt{y}}{z}\right)$ **d** $\log_a\left(\dfrac{x}{a^4}\right)$

a $\log_a (x^2 y z^3)$

$\quad = \log_a (x^2) + \log_a y + \log_a (z^3)$

$\quad = 2\log_a x + \log_a y + 3\log_a z$

b $\log_a\left(\dfrac{x}{y^3}\right)$

$\quad = \log_a x - \log_a (y^3)$

$\quad = \log_a x - 3\log_a y$

c $\log_a\left(\dfrac{x\sqrt{y}}{z}\right)$

$\quad = \log_a (x\sqrt{y}) - \log_a z$

$\quad = \log_a x + \log_a \sqrt{y} - \log_a z$

$\quad = \log_a x + \dfrac{1}{2}\log_a y - \log_a z$

Use the power law ($\sqrt{y} = y^{\frac{1}{2}}$).

d $\log_a\left(\dfrac{x}{a^4}\right)$

$\quad = \log_a x - \log_a (a^4)$

$\quad = \log_a x - 4\log_a a$

$\quad = \log_a x - 4$

$\log_a a = 1$.

Exercise 3D

1 Write as a single logarithm:

   a $\log_2 7 + \log_2 3$

   b $\log_2 36 - \log_2 4$

   c $3\log_5 2 + \log_5 10$

   d $2\log_6 8 - 4\log_6 3$

   e $\log_{10} 5 + \log_{10} 6 - \log_{10}\left(\frac{1}{4}\right)$

2 Write as a single logarithm, then simplify your answer:

   a $\log_2 40 - \log_2 5$

   b $\log_6 4 + \log_6 9$

   c $2\log_{12} 3 + 4\log_{12} 2$

   d $\log_8 25 + \log_8 10 - 3\log_8 5$

   e $2\log_{10} 20 - (\log_{10} 5 + \log_{10} 8)$

3 Write in terms of $\log_a x$, $\log_a y$ and $\log_a z$:

   a $\log_a (x^3 y^4 z)$ **b** $\log_a\left(\dfrac{x^5}{y^2}\right)$

   c $\log_a (a^2 x^2)$ **d** $\log_a\left(\dfrac{x\sqrt{y}}{z}\right)$

   e $\log_a \sqrt{ax}$

3.5 You need to be able to solve equations of the form $a^x = b$.

Example 8

Solve the equation $3^x = 20$, giving your answer to 3 significant figures.

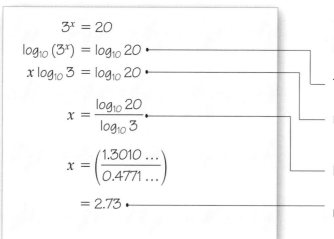

$$3^x = 20$$
$$\log_{10}(3^x) = \log_{10} 20$$
$$x\log_{10} 3 = \log_{10} 20$$
$$x = \frac{\log_{10} 20}{\log_{10} 3}$$
$$x = \left(\frac{1.3010\ldots}{0.4771\ldots}\right)$$
$$= 2.73$$

Since there is no base 3 logarithm button on your calculator, any working must be done using base 10.

Take logs to base 10 on each side.

Use the power law.

Divide by $\log_{10} 3$.

Use your calculator (logs to base 10).

Example 9

Solve the equation $7^{x+1} = 3^{x+2}$, giving your answer to 4 decimal places.

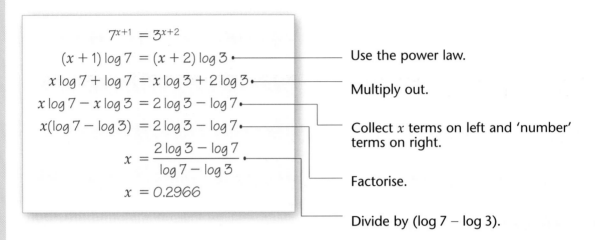

$$7^{x+1} = 3^{x+2}$$

$(x+1)\log 7 = (x+2)\log 3$ ————————— Use the power law.

$x\log 7 + \log 7 = x\log 3 + 2\log 3$ ————— Multiply out.

$x\log 7 - x\log 3 = 2\log 3 - \log 7$

$x(\log 7 - \log 3) = 2\log 3 - \log 7$

———— Collect x terms on left and 'number' terms on right.

$$x = \frac{2\log 3 - \log 7}{\log 7 - \log 3}$$

$$x = 0.2966$$

———— Factorise.

———— Divide by $(\log 7 - \log 3)$.

Example 10

Solve the equation $5^{2x} + 7(5^x) - 30 = 0$, giving your answer to 2 decimal places:

Let $y = 5^x$

$y^2 + 7y - 30 = 0$ ————————— $5^{2x} = (5^x)^2 = y^2$

So $(y+10)(y-3) = 0$

So $\quad y = -10$ or $y = 3$

If $y = -10$, $5^x = -10$ ————— No solution. 5^x cannot be negative.

If $y = 3$, $5^x = 3$

$$\log_{10}(5^x) = \log_{10}3$$

$$x\log_{10}5 = \log_{10}3$$

$$x = \frac{\log_{10}3}{\log_{10}5}$$

———— Solve as in Example 8.

$$x = 0.68 \ (2 \ d.p.)$$

Exercise 3E

1 Solve, giving your answer to 3 significant figures:

a $2^x = 75$ **b** $3^x = 10$

c $5^x = 2$ **d** $4^{2x} = 100$

e $9^{x+5} = 50$ **f** $7^{2x-1} = 23$

g $3^{x-1} = 8^{x+1}$ **h** $2^{2x+3} = 3^{3x+2}$

i $8^{3-x} = 10^x$ **j** $3^{4-3x} = 4^{x+5}$

2 Solve, giving your answer to 3 significant figures:

a $2^{2x} - 6(2^x) + 5 = 0$

b $3^{2x} - 15(3^x) + 44 = 0$

c $5^{2x} - 6(5^x) - 7 = 0$

d $3^{2x} + 3^{x+1} - 10 = 0$

e $7^{2x} + 12 = 7^{x+1}$

> **Hint for question 2d:**
> Note that
> $3^{x+1} = 3^x \times 3^1 = 3(3^x)$

3.6 To evaluate a logarithm using your calculator, you sometimes need to change the base of the logarithm.

Working in base a, suppose that: $\log_a x = m$

Writing this as a power: $a^m = x$

Taking logs to a different base b: $\log_b(a^m) = \log_b x$

Using the power law: $m \log_b a = \log_b x$

Writing m as $\log_a x$: $\log_b x = \log_a x \times \log_b a$

This can be written as:

■ $\log_a x = \dfrac{\log_b x}{\log_b a}$

> **Hint:** This is the change of base rule for logarithms.

Using this rule, notice in particular that $\log_a b = \dfrac{\log_b b}{\log_b a}$, but $\log_b b = 1$, so:

■ $\log_a b = \dfrac{1}{\log_b a}$

Example 11

Find, to 3 significant figures, the value of $\log_8 11$:

$\log_8 11 = \dfrac{\log_{10} 11}{\log_{10} 8}$

$= 1.15$

One method is to use the change of base rule to change to base 10.

$8^x = 11$

$\log_{10}(8^x) = \log_{10} 11$

$x \log_{10} 8 = \log_{10} 11$

$x = \dfrac{\log_{10} 11}{\log_{10} 8}$

$x = 1.15$

Another method is to solve $8^x = 11$.

Take logs to base 10 of each side.

Use the power law.

Divide by $\log_{10} 8$.

Example 12

Solve the equation $\log_5 x + 6 \log_x 5 = 5$:

$\log_5 x + \dfrac{6}{\log_5 x} = 5$ Use change of base rule (special case).

Let $\log_5 x = y$

$y + \dfrac{6}{y} = 5$

$y^2 + 6 = 5y$ Multiply by y.

$y^2 - 5y + 6 = 0$

$(y - 3)(y - 2) = 0$

So $y = 3$ or $y = 2$

$\log_5 x = 3$ or $\log_5 x = 2$

$x = 5^3$ or $x = 5^2$ Write as powers.

$x = 125$ or $x = 25$

Exercise 3F

1 Find, to 3 decimal places:

    a $\log_7 120$ **b** $\log_3 45$ **c** $\log_2 19$

    d $\log_{11} 3$ **e** $\log_6 4$

2 Solve, giving your answer to 3 significant figures:

    a $8^x = 14$ **b** $9^x = 99$ **c** $12^x = 6$

3 Solve, giving your answer to 3 significant figures:

    a $\log_2 x = 8 + 9 \log_x 2$

    b $\log_4 x + 2 \log_x 4 + 3 = 0$

    c $\log_2 x + \log_4 x = 2$

Mixed exercise 3G

1 Find the possible values of x for which $2^{2x+1} = 3(2^x) - 1$.   **E**

2 **a** Express $\log_a (p^2 q)$ in terms of $\log_a p$ and $\log_a q$.

    b Given that $\log_a (pq) = 5$ and $\log_a (p^2 q) = 9$, find the values of $\log_a p$ and $\log_a q$.   **E**

3 Given that $p = \log_q 16$, express in terms of p,

    a $\log_q 2$,

    b $\log_q (8q)$.   **E**

4 a Given that $\log_3 x = 2$, determine the value of x.

 b Calculate the value of y for which $2\log_3 y - \log_3(y + 4) = 2$.

 c Calculate the values of z for which $\log_3 z = 4\log_z 3$. **E**

5 a Using the substitution $u = 2^x$, show that the equation $4^x - 2^{(x+1)} - 15 = 0$ can be written in the form $u^2 - 2u - 15 = 0$.

 b Hence solve the equation $4^x - 2^{(x+1)} - 15 = 0$, giving your answer to 2 decimal places. **E**

6 Solve, giving your answers as exact fractions, the simultaneous equations:

$$8^y = 4^{2x+3}$$
$$\log_2 y = \log_2 x + 4.$$ **E**

7 Find the values of x for which $\log_3 x - 2\log_x 3 = 1$. **E**

8 Solve the equation

$$\log_3(2 - 3x) = \log_9(6x^2 - 19x + 2).$$ **E**

9 If $xy = 64$ and $\log_x y + \log_y x = \dfrac{5}{2}$, find x and y. **E**

10 Prove that if $a^x = b^y = (ab)^{xy}$, then $x + y = 1$. **E**

11 a Show that $\log_4 3 = \log_2 \sqrt{3}$.

 b Hence or otherwise solve the simultaneous equations:

$$2\log_2 y = \log_4 3 + \log_2 x,$$
$$3^y = 9^x,$$

 given that x and y are positive. **E**

12 a Given that $3 + 2\log_2 x = \log_2 y$, show that $y = 8x^2$.

 b Hence, or otherwise, find the roots α and β, where $\alpha < \beta$, of the equation $3 + 2\log_2 x = \log_2(14x - 3)$.

 c Show that $\log_2 \alpha = -2$.

 d Calculate $\log_2 \beta$, giving your answer to 3 significant figures. **E**

Summary of key points

1 A function $y = a^x$, or $f(x) = a^x$, where a is a constant, is called an exponential function.

2 $\log_a n = x$ means that $a^x = n$, where a is called the base of the logarithm.

3 $\log_a 1 = 0$
$\log_a a = 1$

4 $\log_{10} x$ is sometimes written as $\log x$.

5 The laws of logarithms are
$$\log_a xy = \log_a x + \log_a y \qquad \text{(the multiplication law)}$$
$$\log_a \left(\frac{x}{y}\right) = \log_a x - \log_a y \qquad \text{(the division law)}$$
$$\log_a (x)^k = k \log_a x \qquad \text{(the power law)}$$

6 From the power law,
$$\log_a \left(\frac{1}{x}\right) = -\log_a x$$

7 You can solve an equation such as $a^x = b$ by first taking logarithms (to base 10) of each side.

8 The change of base rule for logarithms can be written as $\log_a x = \dfrac{\log_b x}{\log_b a}$

9 From the change of base rule, $\log_a b = \dfrac{1}{\log_b a}$

4 Coordinate geometry in the (x, y) plane

This chapter shows you how to solve problems involving circles.

4.1 You can find the mid-point of the line joining (x_1, y_1) and (x_2, y_2) by using the formula $\left(\dfrac{x_1 + x_2}{2}, \dfrac{y_1 + y_2}{2}\right)$.

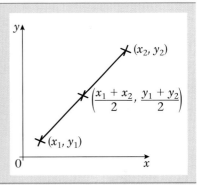

Example 1

Find the mid-point of the line joining these pairs of points:

a $(2, 3), (6, 9)$ **b** $(2a, -4b), (7a, 8b)$ **c** $(4, \sqrt{2}), (-4, 3\sqrt{2})$

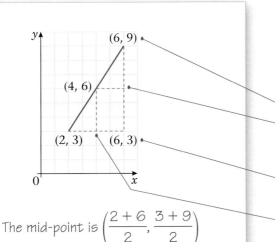

A diagram might help you work this out. The y-coordinate is half way between 3 and 9, so use $\dfrac{y_1 + y_2}{2}$ with $y_1 = 3$ and $y_2 = 9$.

$(6, 3)$ has the same x-coordinate as $(6, 9)$ and the same y-coordinate as $(2, 3)$.

The x-coordinate is half way between 2 and 6, so use $\dfrac{x_1 + x_2}{2}$ with $x_1 = 2$ and $x_2 = 6$.

The mid-point is $\left(\dfrac{2 + 6}{2}, \dfrac{3 + 9}{2}\right)$

Simplify.

$= \left(\dfrac{8}{2}, \dfrac{12}{2}\right)$

$= (4, 6)$

b The mid-point is $\left(\dfrac{2a + 7a}{2}, \dfrac{-4b + 8b}{2}\right)$.

Use $\left(\dfrac{x_1 + x_2}{2}, \dfrac{y_1 + y_2}{2}\right)$.

Here $(x_1, y_1) = (2a, -4b)$ and $(x_2, y_2) = (7a, 8b)$.

$= \left(\dfrac{9a}{2}, \dfrac{4b}{2}\right)$

Simplify.

$= \left(\dfrac{9a}{2}, 2b\right)$

c The mid-point is $\left(\dfrac{4 + (-4)}{2}, \dfrac{\sqrt{2} + 3\sqrt{2}}{2}\right)$.

Use $\left(\dfrac{x_1 + x_2}{2}, \dfrac{y_1 + y_2}{2}\right)$.

Here $(x_1, y_1) = (4, \sqrt{2})$ and $(x_2, y_2) = (-4, 3\sqrt{2})$.

$= \left(\dfrac{4 - 4}{2}, \dfrac{4\sqrt{2}}{2}\right)$

Simplify.

$= (0, 2\sqrt{2})$

Example 2

The line AB is a diameter of a circle, where A and B are $(-3, 8)$ and $(5, 4)$ respectively. Find the coordinates of the centre of the circle.

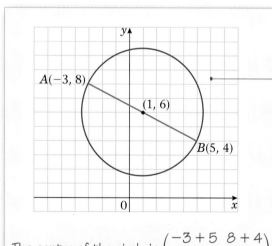

Draw a sketch.

Remember the centre of a circle is the mid-point of a diameter.

The centre of the circle is $\left(\dfrac{-3 + 5}{2}, \dfrac{8 + 4}{2}\right)$.

Use $\left(\dfrac{x_1 + x_2}{2}, \dfrac{y_1 + y_2}{2}\right)$.

Here $(x_1, y_1) = (-3, 8)$ and $(x_2, y_2) = (5, 4)$.

$= \left(\dfrac{2}{2}, \dfrac{12}{2}\right)$

$= (1, 6)$

Example 3

The line PQ is a diameter of the circle centre $(2, -2)$. Given P is $(8, -5)$, find the coordinates of Q.

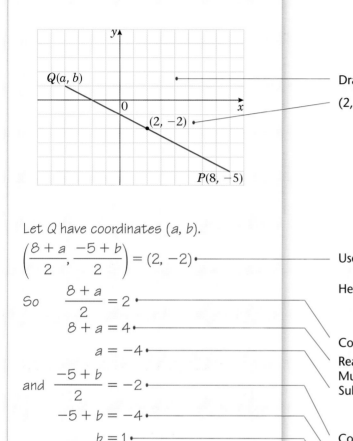

Draw a sketch.

$(2, -2)$ is the mid-point of (a, b) and $(8, -5)$.

Let Q have coordinates (a, b).

$$\left(\frac{8 + a}{2}, \frac{-5 + b}{2}\right) = (2, -2)$$

Use $\left(\frac{x_1 + x_2}{2}, \frac{y_1 + y_2}{2}\right)$.

Here $(x_1, y_1) = (8, -5)$ and $(x_2, y_2) = (a, b)$.

So $\quad \dfrac{8 + a}{2} = 2$

$8 + a = 4$

$a = -4$

Compare the x-coordinates.
Rearrange the equation to find a.
Multiply each side by 2 to clear the fraction.
Subtract 8 from each side.

and $\quad \dfrac{-5 + b}{2} = -2$

$-5 + b = -4$

$b = 1$

Compare the y-coordinates.
Rearrange the equation to find b.
Multiply each side by 2 to clear the fraction.
Add 5 to each side.

So, Q is $(-4, 1)$.

Exercise 4A

1 Find the mid-point of the line joining these pairs of points:

a $(4, 2), (6, 8)$ **b** $(0, 6), (12, 2)$ **c** $(2, 2), (-4, 6)$

d $(-6, 4), (6, -4)$ **e** $(-5, 3), (7, 5)$ **f** $(7, -4), (-3, 6)$

g $(-5, -5), (-11, 8)$ **h** $(6a, 4b), (2a, -4b)$ **i** $(2p, -q), (4p, 5q)$

j $(-2s, -7t), (5s, t)$ **k** $(-4u, 0), (3u, -2v)$ **l** $(a + b, 2a - b), (3a - b, -b)$

m $(4\sqrt{2}, 1), (2\sqrt{2}, 7)$ **n** $(-\sqrt{3}, 3\sqrt{5}), (5\sqrt{3}, 2\sqrt{5})$

o $(\sqrt{2} - \sqrt{3}, 3\sqrt{2} + 4\sqrt{3}), (3\sqrt{2} + \sqrt{3}, -\sqrt{2} + 2\sqrt{3})$

2 The line PQ is a diameter of a circle, where P and Q are $(-4, 6)$ and $(7, 8)$ respectively. Find the coordinates of the centre of the circle.

3 The line RS is a diameter of a circle, where R and S are $\left(\dfrac{4a}{5}, -\dfrac{3b}{4}\right)$ and $\left(\dfrac{2a}{5}, \dfrac{5b}{4}\right)$ respectively. Find the coordinates of the centre of the circle.

4 The line AB is a diameter of a circle, where A and B are $(-3, -4)$ and $(6, 10)$ respectively. Show that the centre of the circle lies on the line $y = 2x$.

5 The line JK is a diameter of a circle, where J and K are $(\tfrac{3}{4}, \tfrac{4}{3})$ and $(-\tfrac{1}{2}, 2)$ respectively. Show that the centre of the circle lies on the line $y = 8x + \tfrac{2}{3}$.

6 The line AB is a diameter of a circle, where A and B are $(0, -2)$ and $(6, -5)$ respectively. Show that the centre of the circle lies on the line $x - 2y - 10 = 0$.

7 The line FG is a diameter of the circle centre $(6, 1)$. Given F is $(2, -3)$, find the coordinates of G.

8 The line CD is a diameter of the circle centre $(-2a, 5a)$. Given D has coordinates $(3a, -7a)$, find the coordinates of C.

9 The points $M(3, p)$ and $N(q, 4)$ lie on the circle centre $(5, 6)$. The line MN is a diameter of the circle. Find the value of p and q.

10 The points $V(-4, 2a)$ and $W(3b, -4)$ lie on the circle centre $(b, 2a)$. The line VW is a diameter of the circle. Find the value of a and b.

Example **4**

The line AB is a diameter of the circle centre C, where A and B are $(-1, 4)$ and $(5, 2)$ respectively. The line l passes through C and is perpendicular to AB. Find the equation of l.

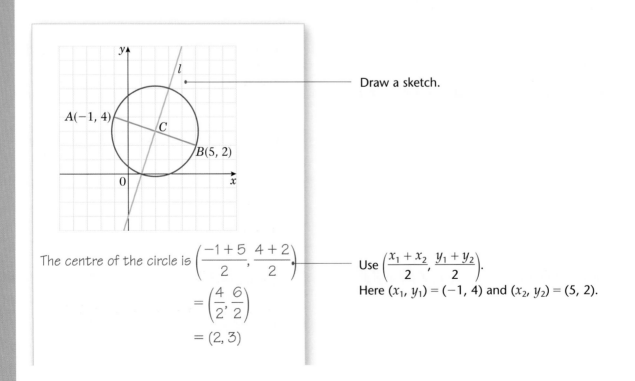

Draw a sketch.

The centre of the circle is $\left(\dfrac{-1+5}{2}, \dfrac{4+2}{2}\right)$ Use $\left(\dfrac{x_1 + x_2}{2}, \dfrac{y_1 + y_2}{2}\right)$.

Here $(x_1, y_1) = (-1, 4)$ and $(x_2, y_2) = (5, 2)$.

$= \left(\dfrac{4}{2}, \dfrac{6}{2}\right)$

$= (2, 3)$

The gradient of the line AB is $\dfrac{2-4}{5-(-1)}$

$$= \dfrac{-2}{6}$$

$$= -\dfrac{1}{3}$$

So, the gradient of the line perpendicular to AB is 3.

The equation of the perpendicular line l is

$$y - 3 = 3(x - 2)$$
$$y - 3 = 3x - 6$$
So $\qquad y = 3x - 3$

Use $m = \dfrac{y_2 - y_1}{x_2 - x_1}$. Here $(x_1, y_1) = (-1, 4)$ and $(x_2, y_2) = (5, 2)$.

Simplify the fraction so divide by 2.

$\dfrac{-1}{3}$ is the same as $-\dfrac{1}{3}$.

Remember the product of the gradients of two perpendicular lines $= -1$, so $-\dfrac{1}{3} \times 3 = -1$.

The perpendicular line l passes through the point (2, 3) and has gradient 3, so use $y - y_1 = m(x - x_1)$ with $m = 3$ and $(x_1, y_1) = (2, 3)$.

Rearrange the equation into the form $y = mx + c$.
Expand the brackets.
Add 3 to each side.

Example 5

The line PQ is a chord of the circle centre $(-3, 5)$, where P and Q are (5, 4) and (1, 12) respectively. The line l is perpendicular to PQ and bisects it. Show that l passes through the centre of the circle.

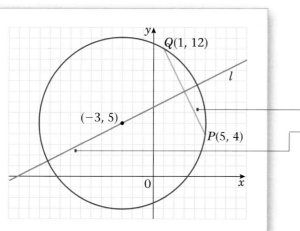

A chord is a line that joins any two points on the circumference of a circle.

The line l **bisects** PQ, so it passes through the mid-point of PQ.

First find the equation of l.

The mid-point of PQ is $\left(\dfrac{5+1}{2}, \dfrac{4+12}{2}\right)$

$$= (3, 8)$$

The gradient of the chord PQ is $\dfrac{12-4}{1-5}$

$$= \dfrac{8}{-4}$$

$$= -2$$

Use $\left(\dfrac{x_1 + x_2}{2}, \dfrac{y_1 + y_2}{2}\right)$.
Here $(x_1, y_1) = (5, 4)$ and $(x_2, y_2) = (1, 12)$.

Use $m = \dfrac{y_2 - y_1}{x_2 - x_1}$.
Here $(x_1, y_1) = (5, 4)$ and $(x_2, y_2) = (1, 12)$.

So, the gradient of the line perpendicular

to PQ is $\dfrac{1}{2}$.

Use the product of the gradients of two perpendicular lines $= -1$, so $-2 \times \frac{1}{2} = -1$.

The equation of the perpendicular line is

$$y - 8 = \frac{1}{2}(x - 3)$$

The perpendicular line passes through the point $(3, 8)$ and has gradient $\frac{1}{2}$, so use $y - y_1 = m(x - x_1)$ with $m = \frac{1}{2}$ and $(x_1, y_1) = (3, 8)$.

$$y - 8 = \frac{1}{2}x - \frac{3}{2}$$

Expand the brackets.

$$y = \frac{1}{2}x - \frac{3}{2} + 8$$

Rearrange the equation into the form $y = mx + c$ by adding 8 to each side.

$$y = \frac{1}{2}x - \frac{3}{2} + \frac{16}{2}$$

$$y = \frac{1}{2}x + \frac{16 - 3}{2}$$

$$y = \frac{1}{2}x + \frac{13}{2}$$

Substitute $x = -3$.

$$y = \frac{1}{2}(-3) + \frac{13}{2}$$

See if $(-3, 5)$ satisfies the equation of l, so substitute the x-coordinate into $y = \frac{1}{2}x + \frac{13}{2}$.

Simplify.

$$= -\frac{3}{2} + \frac{13}{2}$$

$$= \frac{-3 + 13}{2}$$

$$= \frac{10}{2}$$

$$= 5$$

This is the y-coordinate, so $(-3, 5)$ is on the line.

So l passes through $(-3, 5)$.

So l passes through the centre of the circle.

The above example is a particular instance of this circle theorem:

- **The perpendicular from the centre of a circle to a chord bisects the chord.**

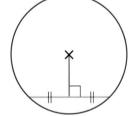

- A line that is perpendicular to a given line and bisects it is called the **perpendicular bisector.**

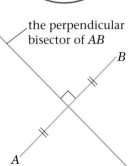

the perpendicular bisector of AB

B

A

Example 6

The lines *AB* and *CD* are chords of a circle. The line $y = 3x - 11$ is the perpendicular bisector of *AB*. The line $y = -x - 1$ is the perpendicular bisector of *CD*. Find the coordinates of the centre of the circle.

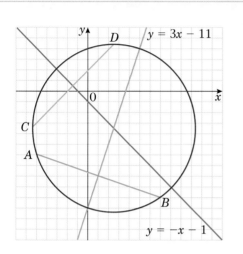

Draw a sketch.

The perpendicular bisectors meet at the centre of the circle.

$$y = 3x - 11$$
$$y = -x - 1$$

Find where the perpendicular bisectors meet, so solve the equations simultaneously.

So $\quad 3x - 11 = -x - 1$

Substitute $y = 3x - 11$ into $y = -x - 1$.

$$4x - 11 = -1$$

Add *x* to each side.

$$4x = 10$$

Add 11 to each side.

$$x = \frac{10}{4}$$

Divide each side by 4.

$$= \frac{5}{2}$$

Simplify the fraction so divide by 2.

Substitute the *x*-coordinate into one of the equations to find the *y*-coordinate.

Sub. $\quad x = \frac{5}{2}$

Here $y = 3x - 11$ and $x = \frac{5}{2}$.

$$y = 3\left(\frac{5}{2}\right) - 11$$

Simplify.

$$= \frac{15}{2} - 11$$

$$= \frac{15}{2} - \frac{22}{2}$$

$$= \frac{15 - 22}{2}$$

$$= -\frac{7}{2}$$

The centre of the circle is $\left(\dfrac{5}{2}, -\dfrac{7}{2}\right)$.

Exercise 4B

1 The line *FG* is a diameter of the circle centre *C*, where *F* and *G* are (−2, 5) and (2, 9) respectively. The line *l* passes through *C* and is perpendicular to *FG*. Find the equation of *l*.

2 The line *JK* is a diameter of the circle centre *P*, where *J* and *K* are (0, −3) and (4, −5) respectively. The line *l* passes through *P* and is perpendicular to *JK*. Find the equation of *l*. Write your answer in the form $ax + by + c = 0$, where *a*, *b* and *c* are integers.

3 The line *AB* is a diameter of the circle centre (4, −2). The line *l* passes through *B* and is perpendicular to *AB*. Given that *A* is (−2, 6),

  a find the coordinates of *B*.

  b Hence, find the equation of *l*.

4 The line *PQ* is a diameter of the circle centre (−4, −2). The line *l* passes through *P* and is perpendicular to *PQ*. Given that *Q* is (10, 4), find the equation of *l*.

5 The line *RS* is a chord of the circle centre (5, −2), where *R* and *S* are (2, 3) and (10, 1) respectively. The line *l* is perpendicular to *RS* and bisects it. Show that *l* passes through the centre of the circle.

6 The line *MN* is a chord of the circle centre (1, $-\frac{1}{2}$), where *M* and *N* are (−5, −5) and (7, 4) respectively. The line *l* is perpendicular to *MN* and bisects it. Find the equation of *l*. Write your answer in the form $ax + by + c = 0$, where *a*, *b* and *c* are integers.

7 The lines *AB* and *CD* are chords of a circle. The line $y = 2x + 8$ is the perpendicular bisector of *AB*. The line $y = -2x - 4$ is the perpendicular bisector of *CD*. Find the coordinates of the centre of the circle.

8 The lines *EF* and *GH* are chords of a circle. The line $y = 3x - 24$ is the perpendicular bisector of *EF*. Given *G* and *F* are (−2, 4) and (4, 10) respectively, find the coordinates of the centre of the circle.

9 The points *P*(3, 16), *Q*(11, 12) and *R*(−7, 6) lie on the circumference of a circle.

  a Find the equation of the perpendicular bisector of

    i *PQ*

    ii *PR*.

  b Hence, find the coordinates of the centre of the circle.

10 The points *A*(−3, 19), *B*(9, 11) and *C*(−15, 1) lie on the circumference of a circle. Find the coordinates of the centre of the circle.

4.2 **You can find the distance *d* between (x_1, y_1) and**

(x_2, y_2) by using the formula

$$d = \sqrt{[(x_2 - x_1)^2 + (y_2 - y_1)^2]}.$$

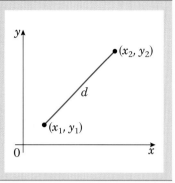

Example 7

Find the distance between these pairs of points:

a $(2, 3), (5, 7)$ **b** $(4a, a), (-3a, 2a)$ **c** $(2\sqrt{2}, -5\sqrt{2}), (4\sqrt{2}, \sqrt{2})$

a

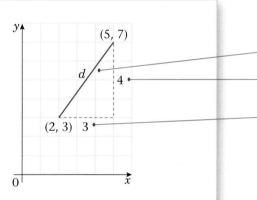

Draw a sketch.

Let the distance between the points be *d*.

The difference in the *y*-coordinates is $7 - 3 = 4$.

The difference in the *x*-coordinates is $5 - 2 = 3$.

$$d^2 = (5 - 2)^2 + (7 - 3)^2$$
$$d^2 = 3^2 + 4^2$$
$$d = \sqrt{(3^2 + 4^2)}$$
$$= \sqrt{25}$$
$$= 5$$

Use Pythagoras' theorem:
$d^2 = (x_2 - x_1)^2 + (y_2 - y_1)^2$

Take the square root of each side.

This is $d = \sqrt{[(x_2 - x_1)^2 + (y_2 - y_1)^2]}$ with $(x_1, y_1) = (2, 3)$ and $(x_2, y_2) = (5, 7)$.

b $d = \sqrt{[(-3a - 4a)^2 + (2a - a)^2]}$
$$= \sqrt{[(-7a)^2 + a^2]}$$
$$= (49a^2 + a^2)$$
$$= 50a^2$$
$$= \sqrt{25 \times 2 \times a^2}$$
$$= \sqrt{25} \times \sqrt{2} \times \sqrt{a^2}$$
$$= 5\sqrt{2}a$$

Use $d = \sqrt{[(x_2 - x_1)^2 + (y_2 - y_1)^2]}$. Here $(x_1, y_1) = (4a, a)$ and $(x_2, y_2) = (-3a, 2a)$.

$(-7a)^2 = -7a \times -7a$
$\qquad = 49a^2$

Simplify.

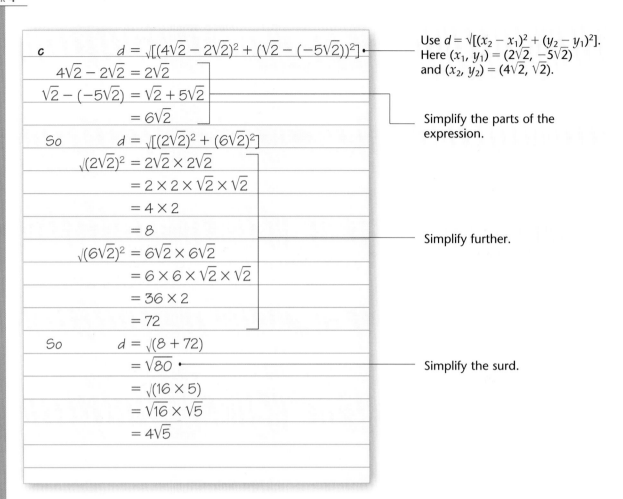

c $\qquad d = \sqrt{[(4\sqrt{2} - 2\sqrt{2})^2 + (\sqrt{2} - (-5\sqrt{2}))^2]}$

$4\sqrt{2} - 2\sqrt{2} = 2\sqrt{2}$

$\sqrt{2} - (-5\sqrt{2}) = \sqrt{2} + 5\sqrt{2}$

$\qquad\qquad = 6\sqrt{2}$

So $\qquad d = \sqrt{[(2\sqrt{2})^2 + (6\sqrt{2})^2]}$

$\sqrt{(2\sqrt{2})^2} = 2\sqrt{2} \times 2\sqrt{2}$

$\qquad\qquad = 2 \times 2 \times \sqrt{2} \times \sqrt{2}$

$\qquad\qquad = 4 \times 2$

$\qquad\qquad = 8$

$\sqrt{(6\sqrt{2})^2} = 6\sqrt{2} \times 6\sqrt{2}$

$\qquad\qquad = 6 \times 6 \times \sqrt{2} \times \sqrt{2}$

$\qquad\qquad = 36 \times 2$

$\qquad\qquad = 72$

So $\qquad d = \sqrt{(8 + 72)}$

$\qquad\qquad = \sqrt{80}$

$\qquad\qquad = \sqrt{(16 \times 5)}$

$\qquad\qquad = \sqrt{16} \times \sqrt{5}$

$\qquad\qquad = 4\sqrt{5}$

Use $d = \sqrt{[(x_2 - x_1)^2 + (y_2 - y_1)^2]}$.
Here $(x_1, y_1) = (2\sqrt{2}, -5\sqrt{2})$
and $(x_2, y_2) = (4\sqrt{2}, \sqrt{2})$.

Simplify the parts of the expression.

Simplify further.

Simplify the surd.

Example 8

The line *PQ* is a diameter of a circle, where *P* and *Q* are $(-1, 3)$ and $(6, -3)$ respectively. Find the radius of the circle.

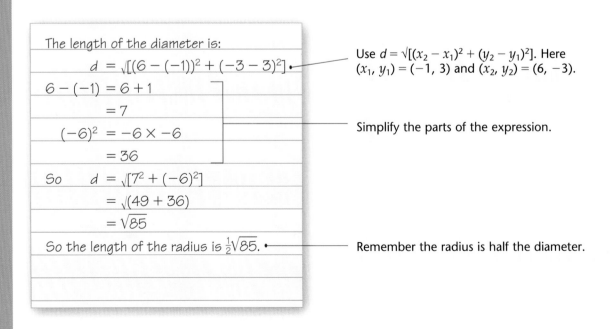

The length of the diameter is:

$\qquad d = \sqrt{[(6 - (-1))^2 + (-3 - 3)^2]}$

$6 - (-1) = 6 + 1$

$\qquad\qquad = 7$

$(-6)^2 = -6 \times -6$

$\qquad\qquad = 36$

So $\qquad d = \sqrt{[7^2 + (-6)^2]}$

$\qquad\qquad = \sqrt{(49 + 36)}$

$\qquad\qquad = \sqrt{85}$

So the length of the radius is $\frac{1}{2}\sqrt{85}$.

Use $d = \sqrt{[(x_2 - x_1)^2 + (y_2 - y_1)^2]}$. Here $(x_1, y_1) = (-1, 3)$ and $(x_2, y_2) = (6, -3)$.

Simplify the parts of the expression.

Remember the radius is half the diameter.

Example 9

The line AB is a diameter of the circle, where A and B are $(-3, 21)$ and $(7, -3)$ respectively. The point $C(14, 4)$ lies on the circumference of the circle. Find the value of AB^2, AC^2 and BC^2. Hence, show that $\angle ACB = 90°$.

$AB^2 = (7 - (-3))^2 + (-3 - 21)^2$ ⟶ Use $d = \sqrt{[(x_2 - x_1)^2 + (y_2 - y_1)^2]}$. Square each side so that $d^2 = (x_2 - x_1)^2 + (y_2 - y_1)^2$. Here $(x_1, y_1) = (-3, 21)$ and $(x_2, y_2) = (7, -3)$.

$\quad = 10^2 + (-24)^2$

$\quad = 676$

$AC^2 = (14 - (-3))^2 + (4 - 21)^2$ ⟶ Here $(x_1, y_1) = (-3, 21)$ and $(x_2, y_2) = (14, 4)$.

$\quad = 17^2 + (-17)^2$

$\quad = 578$

$BC^2 = (14 - 7)^2 + (4 - (-3))^2$ ⟶ Here $(x_1, y_1) = (7, -3)$ and $(x_2, y_2) = (14, 4)$.

$\quad = 7^2 + 7^2$

$\quad = 98$

Now, $578 + 98 = 676$ ⟶ Use Pythagoras' theorem to test if the triangle has a right angle. This is $AC^2 + BC^2 = AB^2$.

So, $\angle ACB = 90°$.

This is a particular instance of this circle theorem:

■ **The angle in a semicircle is a right angle.**

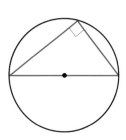

Exercise 4C

1 Find the distance between these pairs of points:

    a $(0, 1)$, $(6, 9)$ **b** $(4, -6)$, $(9, 6)$ **c** $(3, 1)$, $(-1, 4)$

    d $(3, 5)$, $(4, 7)$ **e** $(2, 9)$, $(4, 3)$ **f** $(0, -4)$, $(5, 5)$

    g $(-2, -7)$, $(5, 1)$ **h** $(-4a, 0)$, $(3a, -2a)$ **i** $(-b, 4b)$, $(-4b, -2b)$

    j $(2c, c)$, $(6c, 4c)$ **k** $(-4d, d)$, $(2d, -4d)$ **l** $(-e, -e)$, $(-3e, -5e)$

    m $(3\sqrt{2}, 6\sqrt{2})$, $(2\sqrt{2}, 4\sqrt{2})$ **n** $(-\sqrt{3}, 2\sqrt{3})$, $(3\sqrt{3}, 5\sqrt{3})$

    o $(2\sqrt{3} - \sqrt{2}, \sqrt{5} + \sqrt{3})$, $(4\sqrt{3} - \sqrt{2}, 3\sqrt{5} + \sqrt{3})$

2 The point $(4, -3)$ lies on the circle centre $(-2, 5)$. Find the radius of the circle.

3 The point $(14, 9)$ is the centre of the circle radius 25. Show that $(-10, 2)$ lies on the circle.

4 The line MN is a diameter of a circle, where M and N are $(6, -4)$ and $(0, -2)$ respectively. Find the radius of the circle.

5 The line QR is a diameter of the circle centre C, where Q and R have coordinates $(11, 12)$ and $(-5, 0)$ respectively. The point P is $(13, 6)$.

    a Find the coordinates of C.   **b** Show that P lies on the circle.

6 The points $(-3, 19)$, $(-15, 1)$ and $(9, 1)$ are vertices of a triangle. Show that a circle centre $(-3, 6)$ can be drawn through the vertices of the triangle.

> **Hint for question 6:**
> Show that the vertices are equidistant from the centre of the circle.

7 The line ST is a diameter of the circle c_1, where S and T are $(5, 3)$ and $(-3, 7)$ respectively. The line UV is a diameter of the circle c_2 centre $(4, 4)$. The point U is $(1, 8)$.

  a Find the radius of **i** c_1 **ii** c_2.

  b Find the distance between the centres of c_1 and c_2.

8 The points $U(-2, 8)$, $V(7, 7)$ and $W(-3, -1)$ lie on a circle.

  a Show that $\triangle UVW$ has a right angle.

  b Find the coordinates of the centre of the circle.

9 The points $A(2, 6)$, $B(5, 7)$ and $C(8, -2)$ lie on a circle.

  a Show that $\triangle ABC$ has a right angle.   **b** Find the area of the triangle.

10 The points $A(-1, 9)$, $B(6, 10)$, $C(7, 3)$ and $D(0, 2)$ lie on a circle.

  a Show that $ABCD$ is a square.   **b** Find the area of $ABCD$.

  c Find the centre of the circle.

4.3 **You can write the equation of a circle in the form $(x - a)^2 + (y - b)^2 = r^2$, where (a, b) is the centre and r is the radius.**

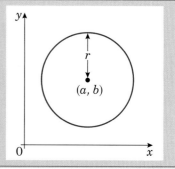

Example 10

Write down the equation of the circle with centre $(5, 7)$ and radius 4.

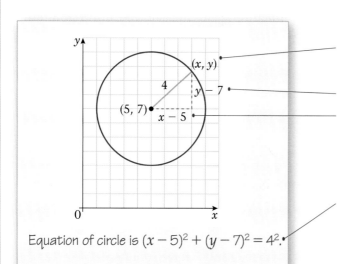

Equation of circle is $(x - 5)^2 + (y - 7)^2 = 4^2$.

(x, y) is *any* point on the circumference of the circle. The distance between (x, y) and $(5, 7)$ is always 4.

The difference in the y-coordinates is $y - 7$.

The difference in the x-coordinates is $x - 5$.

To find the equation of the circle, use $d = \sqrt{[(x_2 - x_1)^2 + (y_2 - y_1)^2]}$. Square each side so that $d^2 = (x_2 - x_1)^2 + (y_2 - y_1)^2$. Here $(x_1, y_1) = (5, 7)$ and $(x_2, y_2) = (x, y)$. This is in the form $(x - a)^2 + (y - b)^2 = r^2$ with $(a, b) = (5, 7)$ and $r = 4$.

Example 11

Write down the coordinates of the centre and the radius of these circles:

a $(x + 3)^2 + (y - 1)^2 = 4^2$ 　　　　　　　**b** $(x - \frac{5}{2})^2 + (y - 3)^2 = 32$

---

a $(x + 3)^2 + (y - 1)^2 = 4^2$

$(x - (-3))^2 + (y - 1)^2 = 4^2$ •——— Write the equation in the form $(x - a)^2 + (y - b)^2 = r^2$, using $-(-3) = +3$.

So centre $= (-3, 1)$, radius $= 4$. •——— So $a = -3$, $b = 1$ and $r = 4$.

---

b $\left(x - \dfrac{5}{2}\right)^2 + (y - 3)^2 = 32$

$\left(x - \dfrac{5}{2}\right)^2 + (y - 3)^2 = (\sqrt{32})^2$ •——— Write the equation in the form $(x - a)^2 + (y - b)^2 = r^2$.

$\sqrt{32} = \sqrt{(16 \times 2)}$ 　　　So $a = \frac{5}{2}$, $b = 3$ and $r = \sqrt{32}$.

$= \sqrt{16} \times \sqrt{2}$ 　　Simplify $\sqrt{32}$.

$= 4\sqrt{2}$

So centre $= \left(\dfrac{5}{2}, 3\right)$, radius $= 4\sqrt{2}$.

Example 12

Show that the circle $(x - 3)^2 + (y + 4)^2 = 20$ passes through $(5, -8)$.

$(x - 3)^2 + (y + 4)^2 = 20$ 　　　　Substitute $x = 5$ and $y = -8$ into the equation of the circle.

Substitute $(5, -8)$ •

$(5 - 3)^2 + (-8 + 4)^2 = 2^2 + (-4)^2$

$= 4 + 16$

$= 20$ •——— $(5, -8)$ satisfies the equation of the circle.

So the circle passes through the point $(5, -8)$.

Example 13

The line AB is a diameter of a circle, where A and B are (4, 7) and (−8, 3) respectively. Find the equation of the circle.

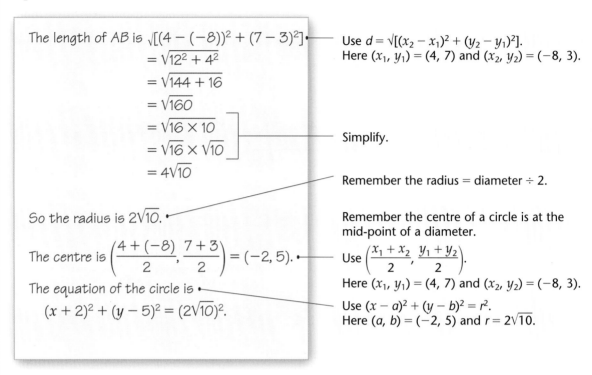

The length of AB is $\sqrt{[(4 - (-8))^2 + (7 - 3)^2]}$ Use $d = \sqrt{[(x_2 - x_1)^2 + (y_2 - y_1)^2]}$.

$$= \sqrt{12^2 + 4^2}$$ Here $(x_1, y_1) = (4, 7)$ and $(x_2, y_2) = (-8, 3)$.

$$= \sqrt{144 + 16}$$

$$= \sqrt{160}$$

$$= \sqrt{16 \times 10}$$ Simplify.

$$= \sqrt{16} \times \sqrt{10}$$

$$= 4\sqrt{10}$$

Remember the radius = diameter ÷ 2.

So the radius is $2\sqrt{10}$.

Remember the centre of a circle is at the mid-point of a diameter.

The centre is $\left(\dfrac{4 + (-8)}{2}, \dfrac{7 + 3}{2}\right) = (-2, 5)$.   Use $\left(\dfrac{x_1 + x_2}{2}, \dfrac{y_1 + y_2}{2}\right)$.

Here $(x_1, y_1) = (4, 7)$ and $(x_2, y_2) = (-8, 3)$.

The equation of the circle is

$$(x + 2)^2 + (y - 5)^2 = (2\sqrt{10})^2.$$

Use $(x - a)^2 + (y - b)^2 = r^2$.

Here $(a, b) = (-2, 5)$ and $r = 2\sqrt{10}$.

Example 14

The line $4x - 3y - 40 = 0$ touches the circle $(x - 2)^2 + (y - 6)^2 = 100$ at $P(10, 0)$. Show that the radius at P is perpendicular to the line.

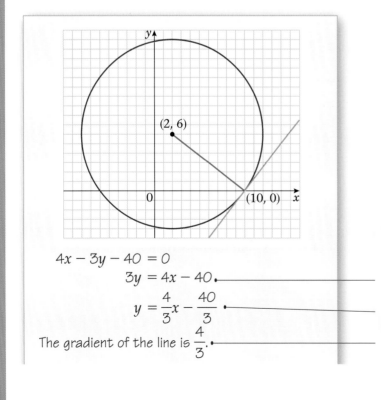

First find the gradient of the line, so rearrange its equation into the form $y = mx + c$.

$$4x - 3y - 40 = 0$$

$$3y = 4x - 40$$

Add 3y to each side and turn the equation around.

$$y = \frac{4}{3}x - \frac{40}{3}$$

Divide each term by 3.

The gradient of the line is $\dfrac{4}{3}$.

Compare $y = \frac{4}{3}x - \frac{40}{3}$ to $y = mx + c$, so $m = \frac{4}{3}$.

$(x - 2)^2 + (y - 6)^2 = 100$

The centre is $(2, 6)$.

So the gradient of the radius at $P = \dfrac{6 - 0}{2 - 10}$

$= \dfrac{6}{-8}$

$= -\dfrac{3}{4}$

Now, $\dfrac{4}{3} \times -\dfrac{3}{4} = -1$.

So, the radius at P is perpendicular to the line.

To find the gradient of the radius at P, first find the centre of the circle from its equation.

Compare $(x - 2)^2 + (y - 6)^2 = 100$ to $(x - a)^2 + (y - b)^2 = r^2$, where (a, b) is the centre.

Use $m = \dfrac{y_2 - y_1}{x_2 - x_1}$.

Here $(x_1, y_1) = (10, 0)$ and $(x_2, y_2) = (2, 6)$.

Simplify the fraction, so divide by 2.

Test to see if the radius is perpendicular to the line.

Use the product of the gradients of two perpendicular lines $= -1$.

The above example is a particular instance of this circle theorem:

■ **The angle between the tangent and a radius is 90°.**

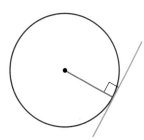

■ **A tangent meets a circle at one point only.**

Exercise 4D

1 Write down the equation of these circles:

   a Centre $(3, 2)$, radius 4

   b Centre $(-4, 5)$, radius 6

   c Centre $(5, -6)$, radius $2\sqrt{3}$

   d Centre $(2a, 7a)$, radius $5a$

   e Centre $(-2\sqrt{2}, -3\sqrt{2})$, radius 1

2 Write down the coordinates of the centre and the radius of these circles:

   a $(x + 5)^2 + (y - 4)^2 = 9^2$

   b $(x - 7)^2 + (y - 1)^2 = 16$

   c $(x + 4)^2 + y^2 = 25$

   d $(x + 4a)^2 + (y + a)^2 = 144a^2$

   e $(x - 3\sqrt{5})^2 + (y + \sqrt{5})^2 = 27$

3 In each case, show that the circle passes through the given point:

    a $(x - 2)^2 + (y - 5)^2 = 13$, $(4, 8)$

    b $(x + 7)^2 + (y - 2)^2 = 65$, $(0, -2)$

    c $x^2 + y^2 = 25^2$, $(7, -24)$

    d $(x - 2a)^2 + (y + 5a)^2 = 20a^2$, $(6a, -3a)$

    e $(x - 3\sqrt{5})^2 + (y - \sqrt{5})^2 = (2\sqrt{10})^2$, $(\sqrt{5}, -\sqrt{5})$

4 The point $(4, -2)$ lies on the circle centre $(8, 1)$. Find the equation of the circle.

5 The line PQ is the diameter of the circle, where P and Q are $(5, 6)$ and $(-2, 2)$ respectively. Find the equation of the circle.

6 The point $(1, -3)$ lies on the circle $(x - 3)^2 + (y + 4)^2 = r^2$. Find the value of r.

7 The line $y = 2x + 13$ touches the circle $x^2 + (y - 3)^2 = 20$ at $(-4, 5)$. Show that the radius at $(-4, 5)$ is perpendicular to the line.

8 The line $x + 3y - 11 = 0$ touches the circle $(x + 1)^2 + (y + 6)^2 = 90$ at $(2, 3)$.

    a Find the radius of the circle.

    b Show that the radius at $(2, 3)$ is perpendicular to the line.

9 The point $P(1, -2)$ lies on the circle centre $(4, 6)$.

    a Find the equation of the circle.

    b Find the equation of the tangent to the circle at P.

10 The tangent to the circle $(x + 4)^2 + (y - 1)^2 = 242$ at $(7, -10)$ meets the y-axis at S and the x-axis at T.

    a Find the coordinates of S and T.

    b Hence, find the area of $\triangle OST$, where O is the origin.

Example **15**

Find where the circle $(x - 5)^2 + (y - 4)^2 = 65$ meets the x-axis.

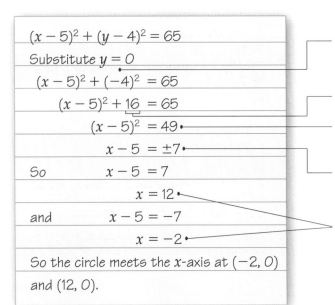

$(x - 5)^2 + (y - 4)^2 = 65$

Substitute $y = 0$

    The circle meets the x-axis when $y = 0$, so substitute $y = 0$ into the equation.

$(x - 5)^2 + (-4)^2 = 65$

$(x - 5)^2 + 16 = 65$

    $-4 \times -4 = 16$

$(x - 5)^2 = 49$

    Subtract 16 from each side.

$x - 5 = \pm 7$

So $x - 5 = 7$

$x = 12$

    Take the square root of each side, so that $\sqrt{49} = \pm 7$.

and $x - 5 = -7$

$x = -2$

    Work out the values of x separately, adding 5 to each side in both cases.

So the circle meets the x-axis at $(-2, 0)$ and $(12, 0)$.

Example 16

Find where the line $y = x + 5$ meets the circle $x^2 + (y - 2)^2 = 29$.

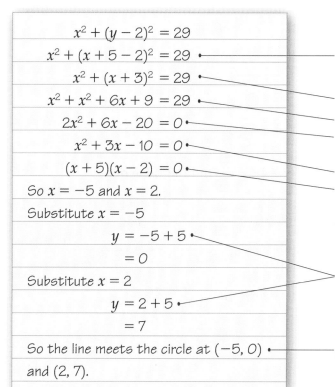

$$x^2 + (y - 2)^2 = 29$$
$$x^2 + (x + 5 - 2)^2 = 29$$
$$x^2 + (x + 3)^2 = 29$$
$$x^2 + x^2 + 6x + 9 = 29$$
$$2x^2 + 6x - 20 = 0$$
$$x^2 + 3x - 10 = 0$$
$$(x + 5)(x - 2) = 0$$

So $x = -5$ and $x = 2$.

Substitute $x = -5$
$$y = -5 + 5$$
$$= 0$$

Substitute $x = 2$
$$y = 2 + 5$$
$$= 7$$

So the line meets the circle at $(-5, 0)$ and $(2, 7)$.

Solve the equations simultaneously, so substitute $y = x + 5$ into the equation of the circle.

Simplify inside the brackets.

Expand the brackets.

Add the x^2 terms and subtract 29 from each side.

Divide each term by 2.

Factorise the quadratic:
$5 \times -2 = -10$
$5 + (-2) = +3$

Now find the y-coordinates, so substitute the values of x into the equation of the line.

Remember to write the answer as coordinates.

Example 17

Show that the line $y = x - 7$ does not meet the circle $(x + 2)^2 + y^2 = 33$.

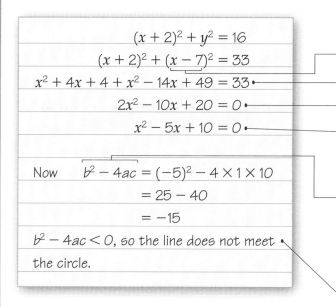

$$(x + 2)^2 + y^2 = 16$$
$$(x + 2)^2 + (x - 7)^2 = 33$$
$$x^2 + 4x + 4 + x^2 - 14x + 49 = 33$$
$$2x^2 - 10x + 20 = 0$$
$$x^2 - 5x + 10 = 0$$

Now
$$b^2 - 4ac = (-5)^2 - 4 \times 1 \times 10$$
$$= 25 - 40$$
$$= -15$$

$b^2 - 4ac < 0$, so the line does not meet the circle.

Solve the equations simultaneously, so substitute $y = x - 7$ into the equation of the circle.

Expand the brackets.

Collect like terms and subtract 33 from each side.

Simplify the quadratic, so divide each term by 2.

Use the discriminant $b^2 - 4ac$ to test for roots of the quadratic equation.

Remember
If $b^2 - 4ac > 0$ there are two distinct roots.
If $b^2 - 4ac = 0$ there is a repeated root.
If $b^2 - 4ac < 0$ there are no real roots.

As $b^2 - 4ac < 0$, there is no solution to the quadratic equation. So, the line does not meet the circle.

Exercise 4E

1 Find where the circle $(x - 1)^2 + (y - 3)^2 = 45$ meets the x-axis.

2 Find where the circle $(x - 2)^2 + (y + 3)^2 = 29$ meets the y-axis.

3 The circle $(x - 3)^2 + (y + 3)^2 = 34$ meets the x-axis at $(a, 0)$ and the y-axis at $(0, b)$. Find the possible values of a and b.

4 The line $y = x + 4$ meets the circle $(x - 3)^2 + (y - 5)^2 = 34$ at A and B. Find the coordinates of A and B.

5 Find where the line $x + y + 5 = 0$ meets the circle $(x + 3)^2 + (y + 5)^2 = 65$.

6 Show that the line $y = x - 10$ does not meet the circle $(x - 2)^2 + y^2 = 25$.

7 Show that the line $x + y = 11$ is a tangent to the circle $x^2 + (y - 3)^2 = 32$.

> **Hint for question 7:**
> Show that the line meets the circle at one point only.

8 Show that the line $3x - 4y + 25 = 0$ is a tangent to the circle $x^2 + y^2 = 25$.

9 The line $y = 2x - 2$ meets the circle $(x - 2)^2 + (y - 2)^2 = 20$ at A and B.

    a Find the coordinates of A and B.

    b Show that AB is a diameter of the circle.

10 The line $x + y = a$ meets the circle $(x - p)^2 + (y - 6)^2 = 20$ at $(3, 10)$, where a and p are constants.

    a Work out the value of a.   **b** Work out the two possible values of p.

Mixed exercise 4F

1 The line $y = 2x - 8$ meets the coordinate axes at A and B. The line AB is a diameter of the circle. Find the equation of the circle.

2 The circle centre $(8, 10)$ meets the x-axis at $(4, 0)$ and $(a, 0)$.

    a Find the radius of the circle.   **b** Find the value of a.

3 The circle $(x - 5)^2 + y^2 = 36$ meets the x-axis at P and Q. Find the coordinates of P and Q.

4 The circle $(x + 4)^2 + (y - 7)^2 = 121$ meets the y-axis at $(0, m)$ and $(0, n)$. Find the value of m and n.

5 The line $y = 0$ is a tangent to the circle $(x - 8)^2 + (y - a)^2 = 16$. Find the value of a.

6 The point $A(-3, -7)$ lies on the circle centre $(5, 1)$. Find the equation of the tangent to the circle at A.

7 The circle $(x + 3)^2 + (y + 8)^2 = 100$ meets the positive coordinate axes at $A(a, 0)$ and $B(0, b)$.

    a Find the value of a and b.

    b Find the equation of the line AB.

8 The circle $(x + 2)^2 + (y - 5)^2 = 169$ meets the positive coordinate axes at $C(c, 0)$ and $D(0, d)$.

    a Find the value of c and d.   **b** Find the area of $\triangle OCD$, where O is the origin.

9 The circle, centre (p, q) radius 25, meets the x-axis at $(-7, 0)$ and $(7, 0)$, where $q > 0$.

    a Find the value of p and q.

    b Find the coordinates of the points where the circle meets the y-axis.

10 Show that $(0, 0)$ lies inside the circle $(x - 5)^2 + (y + 2)^2 = 30$.

11 The points $A(-4, 0)$, $B(4, 8)$ and $C(6, 0)$ lie on a circle. The lines AB and BC are chords of the circle. Find the coordinates of the centre of the circle.

12 The points $R(-4, 3)$, $S(7, 4)$ and $T(8, -7)$ lie on a circle.

    a Show that $\triangle RST$ has a right angle.   **b** Find the equation of the circle.

13 The points $A(-7, 7)$, $B(1, 9)$, $C(3, 1)$ and $D(-7, 1)$ lie on a circle. The lines AB and CD are chords of the circle.

    a Find the equation of the perpendicular bisector of **i** AB **ii** CD.

    b Find the coordinates of the centre of the circle.

14 The centres of the circles $(x - 8)^2 + (y - 8)^2 = 117$ and $(x + 1)^2 + (y - 3)^2 = 106$ are P and Q respectively.

    a Show that P lies on $(x + 1)^2 + (y - 3)^2 = 106$.

    b Find the length of PQ.

15 The line $y = -3x + 12$ meets the coordinate axes at A and B.

    a Find the coordinates of A and B.

    b Find the coordinates of the mid-point of AB.

    c Find the equation of the circle that passes through A, B and O, where O is the origin.

16 The points $A(-5, 5)$, $B(1, 5)$, $C(3, 3)$ and $D(3, -3)$ lie on a circle. Find the equation of the circle.

17 The line AB is a chord of a circle centre $(2, -1)$, where A and B are $(3, 7)$ and $(-5, 3)$ respectively. AC is a diameter of the circle. Find the area of $\triangle ABC$.

18 The points $A(-1, 0)$, $B(\frac{1}{2}, \sqrt{\frac{3}{2}})$ and $C(\frac{1}{2}, -\sqrt{\frac{3}{2}})$ are the vertices of a triangle.

    a Show that the circle $x^2 + y^2 = 1$ passes through the vertices of the triangle.

    b Show that $\triangle ABC$ is equilateral.

19 The points $P(2, 2)$, $Q(2 + \sqrt{3}, 5)$ and $R(2 - \sqrt{3}, 5)$ lie on the circle $(x - 2)^2 + (y - 4)^2 = r^2$.

    a Find the value of r.   **b** Show that $\triangle PQR$ is equilateral.

20 The points $A(-3, -2)$, $B(-6, 0)$ and $C(p, q)$ lie on a circle centre $(-\frac{5}{2}, 2)$. The line BC is a diameter of the circle.

    a Find the value of p and q.

    b Find the gradient of **i** AB **ii** AC.

    c Show that AB is perpendicular to AC.

Summary of key points

1   The mid-point of (x_1, y_1) and (x_2, y_2) is

$\left(\dfrac{x_1 + x_2}{2}, \dfrac{y_1 + y_2}{2}\right)$.

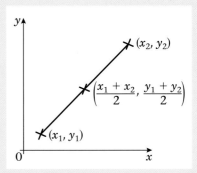

2   The distance d between (x_1, y_1) and (x_2, y_2) is
$d = \sqrt{[(x_2 - x_1)^2 + (y_2 - y_1)^2]}$.

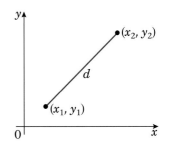

3   The equation of the circle centre (a, b) radius r
is $(x - a)^2 + (y - b)^2 = r^2$.

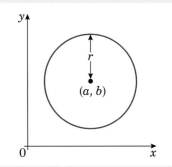

4   A chord is a line that joins two points on the
circumference of a circle.

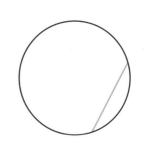

5 The perpendicular from the centre of a circle to a chord bisects the chord.

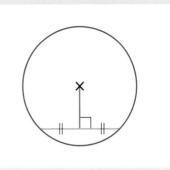

6 The angle in a semicircle is a right angle.

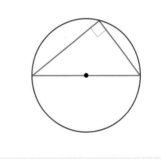

7 A tangent is a line that meets a circle at one point only.

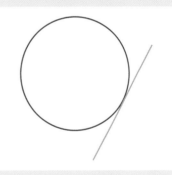

8 The angle between a tangent and a radius is 90°.

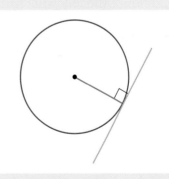

5 The binomial expansion

This chapter shows you how to expand a bracket of the form $(a + b)^n$ for any positive integer n.

5.1 You can use Pascal's Triangle to quickly expand expressions such as $(x + 2y)^3$.

Consider the following:

$(a + b)^1 = a + b$

$(a + b)^2 = (a + b)(a + b) = a^2 + 2ab + b^2$

$(a + b)^3 = (a + b)(a + b)^2 = (a + b)(a^2 + 2ab + b^2)$

$\qquad = a(a^2 + 2ab + b^2) + b(a^2 + 2ab + b^2)$

$\qquad = a^3 + 2a^2b + ab^2 + ba^2 + 2ab^2 + b^3$

$\qquad = a^3 + 3a^2b + 3ab^2 + b^3$

Similarly $(a + b)^4 = a^4 + 4a^3b + 6a^2b^2 + 4ab^3 + b^4$.

Setting these results out in order starting with $(a + b)^0$ we find that:

$$
\begin{array}{llllllllll}
(a + b)^0 = & & & & & 1 & & & \\
(a + b)^1 = & & & & 1a & + & 1b & & \\
(a + b)^2 = & & & 1a^2 & + & 2ab & + & 1b^2 & \\
(a + b)^3 = & & 1a^3 & + & 3a^2b & + & 3ab^2 & + & 1b^3 \\
(a + b)^4 = & 1a^4 & + & 4a^3b & + & 6a^2b^2 & + & 4ab^3 & + & 1b^4
\end{array}
$$

Hint: The terms all have the same index as the original expression. For example, look at the line for $(a + b)^3$. All of the terms have a total index of 3 (a^3, a^2b, ab^2 and b^3).

You should notice the following patterns:

- The coefficients form a pattern that is known as Pascal's Triangle.

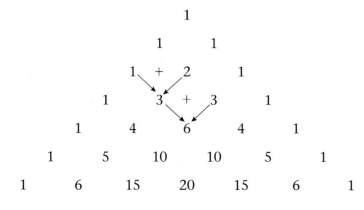

Hint: To get from one line to the next you add adjacent pairs of numbers.

Example 1

Use Pascal's Triangle to find the expansions of:

a $(x + 2y)^3$

b $(2x - 5)^4$

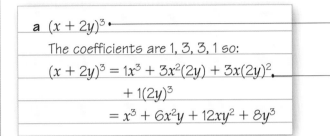

a $(x + 2y)^3$

The coefficients are 1, 3, 3, 1 so:

$(x + 2y)^3 = 1x^3 + 3x^2(2y) + 3x(2y)^2$
$\qquad + 1(2y)^3$
$\qquad = x^3 + 6x^2y + 12xy^2 + 8y^3$

Index = 3 so look at the 4th line in Pascal's Triangle to find the coefficients.

Use the expansion of $(a + b)^3$.
Remember $(2y)^2 = 4y^2$.

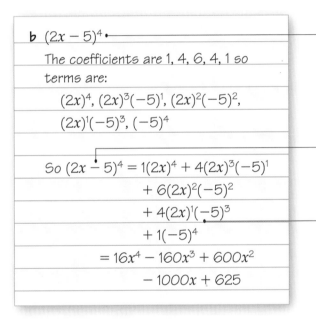

b $(2x - 5)^4$

The coefficients are 1, 4, 6, 4, 1 so terms are:

$(2x)^4, (2x)^3(-5)^1, (2x)^2(-5)^2,$
$(2x)^1(-5)^3, (-5)^4$

So $(2x - 5)^4 = 1(2x)^4 + 4(2x)^3(-5)^1$
$\qquad + 6(2x)^2(-5)^2$
$\qquad + 4(2x)^1(-5)^3$
$\qquad + 1(-5)^4$
$\qquad = 16x^4 - 160x^3 + 600x^2$
$\qquad - 1000x + 625$

Index = 4 so look at the 5th line of Pascal's Triangle.

Use the expansion of $(a + b)^4$.

Careful with the negative numbers!

Example 2

The coefficient of x^2 in the expansion of $(2 - cx)^3$ is 294. Find the possible value(s) of the constant c.

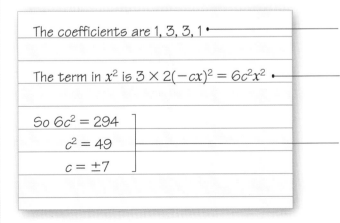

The coefficients are 1, 3, 3, 1

The term in x^2 is $3 \times 2(-cx)^2 = 6c^2x^2$

So $6c^2 = 294$
$c^2 = 49$
$c = \pm 7$

Index = 3, so use the 4th line of Pascal's Triangle to find coefficients.

From the expansion of $(a + b)^3$ the x^2 term is $3ab^2$ where $a = 2$ and $b = -cx$.

Set up and solve an equation in c.

Exercise 5A

1 Write down the expansion of:

   a $(x + y)^4$　　　**b** $(p + q)^5$　　　**c** $(a - b)^3$　　　**d** $(x + 4)^3$

   e $(2x - 3)^4$　　　**f** $(a + 2)^5$　　　**g** $(3x - 4)^4$　　　**h** $(2x - 3y)^4$

2 Find the coefficient of x^3 in the expansion of:

   a $(4 + x)^4$　　　**b** $(1 - x)^5$　　　**c** $(3 + 2x)^3$　　　**d** $(4 + 2x)^5$

   e $(2 + x)^6$　　　**f** $(4 - \frac{1}{2}x)^4$　　　**g** $(x + 2)^5$　　　**h** $(3 - 2x)^4$

3 Fully expand the expression $(1 + 3x)(1 + 2x)^3$.

4 Expand $(2 + y)^3$. Hence or otherwise, write down the expansion of $(2 + x - x^2)^3$ in ascending powers of x.

5 Find the coefficient of the term in x^3 in the expansion of $(2 + 3x)^3(5 - x)^3$.

6 The coefficient of x^2 in the expansion of $(2 + ax)^3$ is 54. Find the possible values of the constant a.

7 The coefficient of x^2 in the expansion of $(2 - x)(3 + bx)^3$ in 45. Find possible values of the constant b.

8 Find the term independent of x in the expansion of $\left(x^2 - \dfrac{1}{2x}\right)^3$.

5.2 **You can use combinations and factorial notation to help you expand binomial expressions. For larger indices, it is quicker than using Pascal's Triangle.**

Suppose that three people A, B and C are running a race. There are six different outcomes for their finishing positions.
The number can be calculated as:

A, B, C
A, C, B
B, A, C
B, C, A
C, A, B
C, B, A

The number can be calculated as:

$3 \times 2 \times 1$

After the first and second places have been awarded, there is only 1 place left for the third place.

There are three runners in the race: A, B or C.

After the winner has crossed the line there are 2 choices for second place.

We can represent $3 \times 2 \times 1$ using what is termed factorial notation.
3!, pronounced '3 factorial' = $3 \times 2 \times 1$.

■ $n! = n \times (n - 1) \times (n - 2) \times (n - 3) \times \ldots \times 3 \times 2 \times 1$

Note: By definition, $0! = 1$

Suppose you wish to choose any two letters from A, B and C, where order does not matter. There are three different outcomes. We can represent this by $^3C_2$ or $\dbinom{3}{2} = \dfrac{3!}{2!1!}$.

■ The number of ways of choosing r items from a group of n items is written ^nC_r or $\binom{n}{r}$ and is calculated by $\dfrac{n!}{(n-r)!r!}$

e.g. $^3C_2 = \dfrac{3!}{(3-2)!2!} = \dfrac{6}{1 \times 2} = 3$

Exercise 5B

1 Find the values of the following:

a $4!$ **b** $6!$ **c** $\dfrac{8!}{6!}$ **d** $\dfrac{10!}{9!}$

e $^4C_2$ **f** $^8C_6$ **g** $^5C_2$ **h** $^6C_3$

i $^{10}C_9$ **j** $^6C_2$ **k** $^8C_5$ **l** $^nC_3$

2 Calculate:

a $^4C_0$ **b** $\binom{4}{1}$ **c** $^4C_2$ **d** $\binom{4}{3}$ **e** $\binom{4}{4}$

Now look at line 4 of Pascal's Triangle. Can you find any connection?

3 Write using combination notation:

a Line 3 of Pascal's Triangle.

b Line 5 of Pascal's Triangle.

4 Why is $^6C_2$ equal to $\binom{6}{4}$?

a Answer using ideas on choosing from a group.

b Answer by calculating both quantities.

5.3 You can use $\binom{n}{r}$ to work out the coefficients in the binomial expansion.

■ The binomial expansion is

$$(a+b)^n = \underbrace{(a+b)(a+b) \ldots (a+b)}_{n \text{ times}}$$

$$= {}^nC_0 a^n + {}^nC_1 a^{n-1}b + {}^nC_2 a^{n-2}b^2 + {}^nC_3 a^{n-3}b^3 + \ldots + {}^nC_n b^n$$

$$\text{or } \binom{n}{0}a^n + \binom{n}{1}a^{n-1}b + \binom{n}{2}a^{n-2}b^2 + \binom{n}{3}a^{n-3}b + \ldots + \binom{n}{n}b^n$$

■ Similarly,

$$(a+bx)^n = {}^nC_0 a^n + {}^nC_1 a^{n-1}bx + {}^nC_2 a^{n-2}b^2x^2 + {}^nC_3 a^{n-3}b^3x^3 + \ldots {}^nC_n b^n x^n$$

$$\text{or } \binom{n}{0}a^n + \binom{n}{1}a^{n-1}bx + \binom{n}{2}a^{n-2}b^2x^2 + \binom{n}{3}a^{n-3}b^3x^3 + \ldots + \binom{n}{n}b^n x^n$$

Hint: You do not need to memorise **both** these forms of the binomial expansion. You should be able to work out this form from the expansion of $(a+b)^n$.

Example 3

Use the binomial theorem to find the expansion of $(3 - 2x)^5$:

$$(3 - 2x)^5 = 3^5 + \binom{5}{1}3^4(-2x) + \binom{5}{2}3^3(-2x)^2$$
$$+ \binom{5}{3}3^2(-2x)^3 + \binom{5}{4}3^1(-2x)^4$$
$$+ (-2x)^5$$
$$= 243 - 810x + 1080x^2 - 720x^3$$
$$+ 240x^4 - 32x^5$$

There will be 6 terms.
The terms have a total index of 5.
Use $(a + bx)^n$ where $a = 3$, $b = -2x$ and $n = 5$.
There are $\binom{5}{2}$ ways of choosing 2 '$-2x$' terms from 5 brackets.

Example 4

Find the first four terms in ascending powers of x of $\left(1 - \dfrac{x}{4}\right)^{10}$ and, by using a suitable

substitution, use your result to find an approximate value to $(0.975)^{10}$. Use your calculator to find the degree of accuracy of your approximation.

$$\left(1 - \frac{x}{4}\right)^{10}$$

Terms are 1^{10}, $1^9\left(-\dfrac{x}{4}\right)^1$, $1^8\left(-\dfrac{x}{4}\right)^2$, and $1^7\left(-\dfrac{x}{4}\right)^3$.

Coefficients are $^{10}C_0$ $^{10}C_1$ $^{10}C_2$ $^{10}C_3$

Combining, we get the first four terms to equal:

$$^{10}C_0 1^{10} + {}^{10}C_1(1)^9\left(-\frac{x}{4}\right)^1 + {}^{10}C_2(1)^8\left(-\frac{x}{4}\right)^2 +$$
$$^{10}C_3(1)^7\left(-\frac{x}{4}\right)^3 + \dots$$
$$= 1 - 2.5x + 2.8125x^2 - 1.875x^3 \dots$$

We want $\left(1 - \dfrac{x}{4}\right) = 0.975$

$$\frac{x}{4} = 0.025$$
$$x = 0.1$$

Substitute $x = 0.1$ into the expansion for $\left(1 - \dfrac{x}{4}\right)^{10}$:

$$0.975^{10} = 1 - 0.25 + 0.028\,125 - 0.001\,875$$
$$= 0.77625$$

Using a calculator $(0.975)^{10} = 0.776\,329\,62$

So approximation is correct to 4 decimal places.

All terms have total index = 10.

You are selecting 2 '$-\dfrac{x}{4}$'s from 10 brackets.

Calculate the value of x.

Substitute $x = 0.1$ into your expansion.

Exercise 5C

1 Write down the expansion of the following:

  a $(2x + y)^4$ **b** $(p - q)^5$ **c** $(1 + 2x)^4$ **d** $(3 + x)^4$
  e $(1 - \frac{1}{2}x)^4$ **f** $(4 - x)^4$ **g** $(2x + 3y)^5$ **h** $(x + 2)^6$

2 Find the term in x^3 of the following expansions:

  a $(3 + x)^5$ **b** $(2x + y)^5$ **c** $(1 - x)^6$ **d** $(3 + 2x)^5$
  e $(1 + x)^{10}$ **f** $(3 - 2x)^6$ **g** $(1 + x)^{20}$ **h** $(4 - 3x)^7$

3 Use the binomial theorem to find the first four terms in the expansion of:

  a $(1 + x)^{10}$ **b** $(1 - 2x)^5$ **c** $(1 + 3x)^6$ **d** $(2 - x)^8$
  e $(2 - \frac{1}{2}x)^{10}$ **f** $(3 - x)^7$ **g** $(x + 2y)^8$ **h** $(2x - 3y)^9$

4 The coefficient of x^2 in the expansion of $(2 + ax)^6$ is 60.
  Find possible values of the constant a.

5 The coefficient of x^3 in the expansion of $(3 + bx)^5$ is -720.
  Find the value of the constant b.

6 The coefficient of x^3 in the expansion of $(2 + x)(3 - ax)^4$ is 30.
  Find the values of the constant a.

7 Write down the first four terms in the expansion of $\left(1 - \dfrac{x}{10}\right)^6$.

  By substituting an appropriate value for x, find an approximate value to $(0.99)^6$. Use your calculator to find the degree of accuracy of your approximation.

8 Write down the first four terms in the expansion of $\left(2 + \dfrac{x}{5}\right)^{10}$.

  By substituting an appropriate value for x, find an approximate value to $(2.1)^{10}$. Use your calculator to find the degree of accuracy of your approximation.

5.4 You need to be able to expand $(1 + x)^n$ and $(a + bx)^n$ using the binomial expansion.

■ $(1 + x)^n = \dbinom{n}{0}1^n + \dbinom{n}{1}1^{n-1}x^1 + \dbinom{n}{2}1^{n-2}x^2 + \dbinom{n}{3}1^{n-3}x^3 + \dbinom{n}{4}1^{n-4}x^4 + \dots + \dbinom{n}{r}1^{n-r}x^r$

$= 1 + nx + \dfrac{n(n-1)}{2!}x^2 + \dfrac{n(n-1)(n-2)}{3!}x^3 + \dfrac{n(n-1)(n-2)(n-3)}{4!}x^4 + \dots$

Example 5

Find the first four terms of the binomial expansion to **a** $(1 + 2x)^5$ and **b** $(2 - x)^6$:

a $(1 + 2x)^5 = 1 + nx + \dfrac{n(n-1)}{2!}x^2 + \dfrac{n(n-1)(n-2)}{3!}x^3 + \ldots$

Compare $(1 + x)^n$ with $(1 + 2x)^n$.

$= 1 + 5(2x) + \dfrac{5(4)}{2!}(2x)^2 + \dfrac{5(4)(3)}{3!}(2x)^3 + \ldots$

Replace n by 5 and 'x' by $2x$.

$= 1 + 10x + 40x^2 + 80x^3 + \ldots$

b $(2 - x)^6 = \left[2\left(1 - \dfrac{x}{2}\right)\right]^6$

The expansion only works for $(1 + x)^n$, so take out a common factor of 2.

$= 2^6\left(1 - \dfrac{x}{2}\right)^6$

$= 2^6\left(1 + 6\left(-\dfrac{x}{2}\right) + \dfrac{6 \times 5}{2!}\left(-\dfrac{x}{2}\right)^2\right.$

Replace n by 6 and 'x' by $-\dfrac{x}{2}$' in the expansion of $(1 + x)^n$.

$\left. + \dfrac{6 \times 5 \times 4}{3!}\left(-\dfrac{x}{2}\right)^3 + \ldots\right)$

$= 2^6\left(1 - 3x + \dfrac{15}{4}x^2 - \dfrac{5}{2}x^3 + \ldots\right)$

Multiply terms in bracket by 2^6.

$= 64 - 192x + 240x^2 - 160x^3 + \ldots$

Exercise 5D

1 Use the binomial expansion to find the first four terms of

a $(1 + x)^8$ **b** $(1 - 2x)^6$ **c** $\left(1 + \dfrac{x}{2}\right)^{10}$

d $(1 - 3x)^5$ **e** $(2 + x)^7$ **f** $(3 - 2x)^3$

g $(2 - 3x)^6$ **h** $(4 + x)^4$ **i** $(2 + 5x)^7$

2 If x is so small that terms of x^3 and higher can be ignored, show that:

$(2 + x)(1 - 3x)^5 \approx 2 - 29x + 165x^2$

3 If x is so small that terms of x^3 and higher can be ignored, and

$(2 - x)(3 + x)^4 \approx a + bx + cx^2$

find the values of the constants a, b and c.

4 When $(1 - 2x)^p$ is expanded, the coefficient of x^2 is 40. Given that $p > 0$, use this information to find:

   a The value of the constant p.

   b The coefficient of x.

   c The coefficient of x^3.

5 Write down the first four terms in the expansion of $(1 + 2x)^8$. By substituting an appropriate value of x (which should be stated), find an approximate value of 1.02^8. State the degree of accuracy of your answer.

Mixed exercise 5E

1 When $(1 - \frac{3}{2}x)^p$ is expanded in ascending powers of x, the coefficient of x is -24.

   a Find the value of p.

   b Find the coefficient of x^2 in the expansion.

   c Find the coefficient of x^3 in the expansion.

2 Given that:

$$(2 - x)^{13} \equiv A + Bx + Cx^2 + \dots$$

Find the values of the integers A, B and C.

3 **a** Expand $(1 - 2x)^{10}$ in ascending powers of x up to and including the term in x^3, simplifying each coefficient in the expansion.

   b Use your expansion to find an approximation to $(0.98)^{10}$, stating clearly the substitution which you have used for x.

4 **a** Use the binomial series to expand $(2 - 3x)^{10}$ in ascending powers of x up to and including the term in x^3, giving each coefficient as an integer.

   b Use your series expansion, with a suitable value for x, to obtain an estimate for 1.97^{10}, giving your answer to 2 decimal places.

5 **a** Expand $(3 + 2x)^4$ in ascending powers of x, giving each coefficient as an integer.

   b Hence, or otherwise, write down the expansion of $(3 - 2x)^4$ in ascending powers of x.

   c Hence by choosing a suitable value for x show that $(3 + 2\sqrt{2})^4 + (3 - 2\sqrt{2})^4$ is an integer and state its value.

6 The coefficient of x^2 in the binomial expansion of $\left(1 + \dfrac{x}{2}\right)^n$, where n is a positive integer, is 7.

   a Find the value of n.

   b Using the value of n found in part **a**, find the coefficient of x^4.

7 **a** Use the binomial theorem to expand $(3 + 10x)^4$ giving each coefficient as an integer.

   b Use your expansion, with an appropriate value for x, to find the exact value of $(1003)^4$. State the value of x which you have used.

8 a Expand $(1 + 2x)^{12}$ in ascending powers of x up to and including the term in x^3, simplifying each coefficient.

b By substituting a suitable value for x, which must be stated, into your answer to part **a**, calculate an approximate value of $(1.02)^{12}$.

c Use your calculator, writing down all the digits in your display, to find a more exact value of $(1.02)^{12}$.

d Calculate, to 3 significant figures, the percentage error of the approximation found in part **b**.

9 Expand $\left(x - \dfrac{1}{x}\right)^5$, simplifying the coefficients.

10 In the binomial expansion of $(2k + x)^n$, where k is a constant and n is a positive integer, the coefficient of x^2 is equal to the coefficient of x^3.

a Prove that $n = 6k + 2$.

b Given also that $k = \frac{2}{3}$, expand $(2k + x)^n$ in ascending powers of x up to and including the term in x^3, giving each coefficient as an exact fraction in its simplest form.

11 a Expand $(2 + x)^6$ as a binomial series in ascending powers of x, giving each coefficient as an integer.

b By making suitable substitutions for x in your answer to part **a**, show that $(2 + \sqrt{3})^6 - (2 - \sqrt{3})^6$ can be simplified to the form $k\sqrt{3}$, stating the value of the integer k.

12 The coefficient of x^2 in the binomial expansion of $(2 + kx)^8$, where k is a positive constant, is 2800.

a Use algebra to calculate the value of k.

b Use your value of k to find the coefficient of x^3 in the expansion.

13 a Given that
$$(2 + x)^5 + (2 - x)^5 \equiv A + Bx^2 + Cx^4,$$
find the value of the constants A, B and C.

b Using the substitution $y = x^2$ and your answers to part **a**, solve
$$(2 + x)^5 + (2 - x)^5 = 349.$$

14 In the binomial expansion of $(2 + px)^5$, where p is a constant, the coefficient of x^3 is 135. Calculate:

a The value of p,

b The value of the coefficient of x^4 in the expansion.

Summary of key points

1 You can use Pascal's Triangle to multiply out a bracket.

2 You can use combinations and factional notation to help you expand binomial expressions. For larger indices it is quicker than using Pascal's Triangle.

3 $n! = n \times (n-1) \times (n-2) \times (n-3) \times \ldots \times 3 \times 2 \times 1$

4 The number of ways of choosing r items from a group of n items is written ^nC_r or $\binom{n}{r}$.

e.g. $^3C_2 = \dfrac{3!}{(3-2)!2!} = \dfrac{6}{1 \times 2} = 3$

5 The binomial expansion is

$$(a+b)^n = {}^nC_0 a^n + {}^nC_1 a^{n-1}b + {}^nC_2 a^{n-2}b + {}^nC_3 a^{n-3}b^3 + \ldots + {}^nC_n b^n$$

or $\binom{n}{0}a^n + \binom{n}{1}a^{n-1}b + \binom{n}{2}a^{n-2}b^2 + \binom{n}{3}a^{n-3}b^3 + \ldots + \binom{n}{n}b^n$

6 Similarly,

$$(a+bx)^n = {}^nC_0 a^n + {}^nC_1 a^{n-1}bx + {}^nC_2 a^{n-2}b^2x^2 + {}^nC_3 a^{n-3}b^3x^3 + \ldots {}^nC_n b^n x^n$$

or $\binom{n}{0}a^n + \binom{n}{1}a^{n-1}bx + \binom{n}{2}a^{n-2}b^2x^2 + \binom{n}{3}a^{n-3}b^3x^3 + \ldots + \binom{n}{n}b^n x^n$

7 $(1+x)^n = 1 + nx + \dfrac{n(n-1)}{2!}x^2 + \dfrac{n(n-1)(n-2)}{3!}x^3 + \dfrac{n(n-1)(n-2)(n-3)}{4!}x^4 + \ldots$

6.1 You can measure angles in radians.

In Chapter 2 you worked with angles in degrees, where one degree is $\frac{1}{360}$th of a complete revolution. This convention dates back to the Babylonians. It has the advantage that 360 has a great number of factors making division of the circle that much easier, but it is still only a convention. Another and perhaps initially stranger measure of an angle is the radian.

■ **If the arc *AB* has length *r*, then ∠AOB is 1 radian (1ᶜ or 1 rad).**

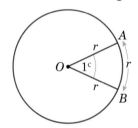

Hint: The symbol for radians is ᶜ, so θ^c means that θ is in radians. If there is no symbol with an angle you should assume that it is in radians, unless the context makes it clear that it is in degrees.

You can put this into words:

■ **A radian is the angle subtended at the centre of a circle by an arc whose length is equal to that of the radius of the circle.**

As an arc of length *r* subtends 1 radian at the centre of the circle, it follows that the circumference (an arc of length $2\pi r$) subtends 2π radians at the centre.

As the circumference subtends an angle of 360° at the centre,

$$2\pi \text{ radians} = 360°$$
$$\text{so } \pi \text{ radians} = 180°$$

It follows that 1 rad = 57.295 …°.

■ **1 radian = $\dfrac{180°}{\pi}$**

Example 1

Convert the following angles into degrees:

a $\dfrac{7\pi}{8}$ rad

b $\dfrac{4\pi}{15}$ rad

a $\dfrac{7\pi}{8}$ rad

$= \dfrac{7}{8} \times 180°$

$= 157.5°$

b $\dfrac{4\pi}{15}$ rad

$= 4 \times \dfrac{180°}{15}$

$= 48°$

Remember that π rad = 180°.
Check using your calculator.

Example 2

Convert the following angles into radians:

a 150° **b** 110°

Caution: If you have been working in radian mode on your calculator make sure you return to degree mode when working with questions involving degrees.

a $150° = 150 \times \dfrac{\pi}{180}$ rad

$= \dfrac{5\pi}{6}$ rad

b $110° = 110 \times \dfrac{\pi}{180}$ rad

$= \dfrac{11}{18}\pi$ rad

Since $180° = \pi$ rad, $1° = \dfrac{\pi}{180}$ rad.

It is worth remembering that $30° = \dfrac{\pi}{6}$ rad.

Your calculator will give the decimal answer 1.919 86 …

These answers, in terms of π, are exact.

Exercise 6A

1 Convert the following angles in radians to degrees:

a $\dfrac{\pi}{20}$ **b** $\dfrac{\pi}{15}$ **c** $\dfrac{5\pi}{12}$

d $\dfrac{\pi}{2}$ **e** $\dfrac{7\pi}{9}$ **f** $\dfrac{7\pi}{6}$

g $\dfrac{5\pi}{4}$ **h** $\dfrac{3\pi}{2}$ **i** 3π

2 Use your calculator to convert the following angles to degrees, giving your answer to the nearest 0.1°:

a 0.46^c **b** 1^c **c** 1.135^c **d** $\sqrt{3}^c$

e 2.5^c **f** 3.14^c **g** 3.49^c

3 Use your calculator to write down the value, to 3 significant figures, of the following trigonometric functions.

a $\sin 0.5^c$ **b** $\cos \sqrt{2}^c$ **c** $\tan 1.05^c$ **d** $\sin 2^c$ **e** $\cos 3.6^c$

4 Convert the following angles to radians, giving your answers as multiples of π:

a 8° **b** 10° **c** 22.5° **d** 30°

e 45° **f** 60° **g** 75° **h** 80°

i 112.5° **j** 120° **k** 135° **l** 200°

m 240° **n** 270° **o** 315° **p** 330°

5 Use your calculator to convert the following angles to radians, giving your answers to 3 significant figures:

a 50° **b** 75° **c** 100°

d 160° **e** 230° **f** 320°.

6.2 **The formula for the length of an arc of a circle is simpler when you use radians.**

■ To find the arc length l of a circle use the formula $l = r\theta$, where r is the radius of the circle and θ is the angle, in radians, contained by the sector.

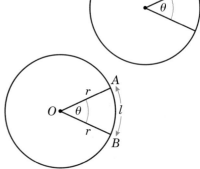

Example 3

Show that the length of an arc is $l = r\theta$.

The circle has centre O and radius r.

The arc AB has length l.

So $\dfrac{l}{2\pi r} = \dfrac{\theta}{2\pi}$.

$\qquad l = r\theta$

$\dfrac{\text{Length of arc}}{\text{Circumference}} = \dfrac{\text{angle } AOB}{\text{total angle around } O}$

(both angles are in radians).

Multiply throughout by $2\pi r$.
If you know two of r, θ and l, the third can be found.

Example 4

Find the length of the arc of a circle of radius 5.2 cm, given that the arc subtends an angle of 0.8 rad at the centre of the circle.

Arc length $= 5.2 \times 0.8$ cm

$\qquad\quad = 4.16$ cm

Use $l = r\theta$, with $r = 5.2$ and $\theta = 0.8$.

Example 5

An arc AB of a circle, with centre O and radius r cm, subtends an angle of θ radians at O. The perimeter of the sector AOB is P cm. Express r in terms of θ.

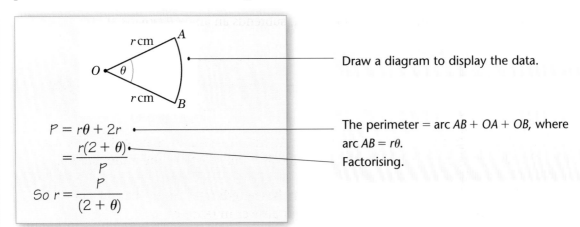

Draw a diagram to display the data.

$P = r\theta + 2r$

$\quad = \dfrac{r(2 + \theta)}{P}$

So $r = \dfrac{P}{(2 + \theta)}$

The perimeter $=$ arc $AB + OA + OB$, where arc $AB = r\theta$.

Factorising.

Example 6

The border of a garden pond consists a straight edge AB of length 2.4 m, and a curved part C, as shown in the diagram below. The curved part is an arc of a circle, centre O and radius 2 m. Find the length of C.

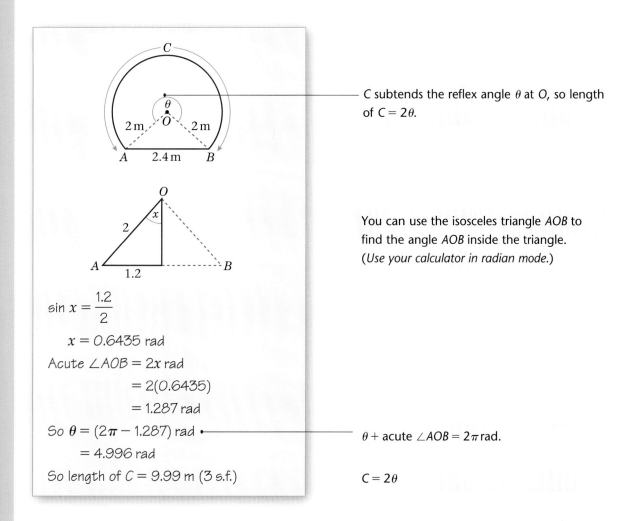

C subtends the reflex angle θ at O, so length of $C = 2\theta$.

You can use the isosceles triangle AOB to find the angle AOB inside the triangle.
(*Use your calculator in radian mode.*)

$\sin x = \dfrac{1.2}{2}$

$x = 0.6435$ rad

Acute $\angle AOB = 2x$ rad

$\qquad\qquad = 2(0.6435)$

$\qquad\qquad = 1.287$ rad

So $\theta = (2\pi - 1.287)$ rad ·

$\qquad = 4.996$ rad

So length of $C = 9.99$ m (3 s.f.)

θ + acute $\angle AOB = 2\pi$ rad.

$C = 2\theta$

Exercise 6B

1. An arc AB of a circle, centre O and radius r cm, subtends an angle θ radians at O. The length of AB is l cm.

   a Find l when   **i** $r = 6$, $\theta = 0.45$ **ii** $r = 4.5$, $\theta = 0.45$ **iii** $r = 20$, $\theta = \frac{3}{8}\pi$

   b Find r when   **i** $l = 10$, $\theta = 0.6$ **ii** $l = 1.26$, $\theta = 0.7$ **iii** $l = 1.5\pi$, $\theta = \frac{5}{12}\pi$

   c Find θ when   **i** $l = 10$, $r = 7.5$ **ii** $l = 4.5$, $r = 5.625$ **iii** $l = \sqrt{12}$, $r = \sqrt{3}$

2. A minor arc AB of a circle, centre O and radius 10 cm, subtends an angle x at O. The major arc AB subtends an angle $5x$ at O. Find, in terms of π, the length of the minor arc AB.

3. An arc AB of a circle, centre O and radius 6 cm, has length l cm. Given that the chord AB has length 6 cm, find the value of l, giving your answer in terms of π.

4 The sector of a circle of radius $\sqrt{10}$ cm contains an angle of $\sqrt{5}$ radians, as shown in the diagram. Find the length of the arc, giving your answer in the form $p\sqrt{q}$ cm, where p and q are integers.

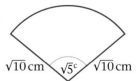

$\sqrt{10}$ cm $\sqrt{5}^c$ $\sqrt{10}$ cm

5 Referring to the diagram, find:

a The perimeter of the shaded region when $\theta = 0.8$ radians.

b The value of θ when the perimeter of the shaded region is 14 cm.

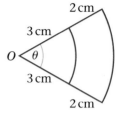

2 cm
3 cm
O θ
3 cm
2 cm

6 A sector of a circle of radius r cm contains an angle of 1.2 radians. Given that the sector has the same perimeter as a square of area 36 cm², find the value of r.

7 A sector of a circle of radius 15 cm contains an angle of θ radians. Given that the perimeter of the sector is 42 cm, find the value of θ.

8 In the diagram AB is the diameter of a circle, centre O and radius 2 cm. The point C is on the circumference such that $\angle COB = \frac{2}{3}\pi$ radians.

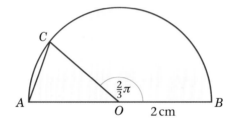

C
$\frac{2}{3}\pi$
A O 2 cm B

a State the value, in radians, of $\angle COA$.

The shaded region enclosed by the chord AC, arc CB and AB is the template for a brooch.

b Find the exact value of the perimeter of the brooch.

9 The points A and B lie on the circumference of a circle with centre O and radius 8.5 cm. The point C lies on the major arc AB. Given that $\angle ACB = 0.4$ radians, calculate the length of the minor arc AB.

10 In the diagram OAB is a sector of a circle, centre O and radius R cm, and $\angle AOB = 2\theta$ radians. A circle, centre C and radius r cm, touches the arc AB at T, and touches OA and OB at D and E respectively, as shown.

a Write down, in terms of R and r, the length of OC.

b Using $\triangle OCE$, show that $R \sin \theta = r (1 + \sin \theta)$.

c Given that $\sin \theta = \frac{3}{4}$ and that the perimeter of the sector OAB is 21 cm, find r, giving your answer to 3 significant figures.

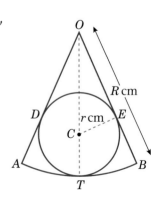

O
R cm
D r cm E
C
A B
T

6.3 **The formula for the area of a sector of a circle is simpler when you use radians.**

■ To find the area A of a sector of a circle use the formula $A = \frac{1}{2}r^2\theta$,
where r is the radius of the circle and θ is the angle, in radians,
contained by the sector.

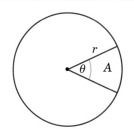

Example 7

Show that the area of the sector of a circle with radius r is $A = \frac{1}{2}r^2\theta$.

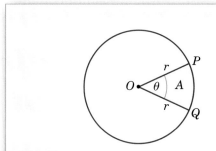

The circle has centre O and radius r.

The sector POQ has area A.

So $\dfrac{A}{\pi r^2} = \dfrac{\theta}{2\pi}$

$A = \dfrac{1}{2}r^2\theta$

$\dfrac{\text{area of sector}}{\text{area of circle}} = \dfrac{\text{angle } POQ}{\text{total angle around } O}$

Multiply throughout by πr^2.
If you know two of r, θ and A, the third can
be found.

Example 8

In the diagram, the area of the minor sector AOB is
$28.9\ \text{cm}^2$.

Given that $\angle AOB = 0.8$ radians, calculate the value
of r.

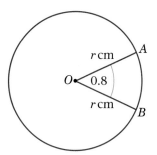

$28.9 = \dfrac{1}{2}r^2(0.8) = 0.4r^2$

So $r^2 = \dfrac{28.9}{0.4} = 72.25$

$r = 8.5$

Let area of sector be $A\ \text{cm}^2$, and use $A = \frac{1}{2}r^2\theta$.

Find r^2 and then take the square root.

Example 9

A plot of land is in the shape of a sector of a circle of radius 55 m. The length of fencing that is erected along the edge of the plot to enclose the land is 176 m. Calculate the area of the plot of land.

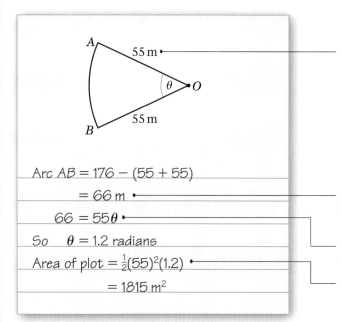

Arc $AB = 176 - (55 + 55)$

$\quad = 66$ m •————————— Draw a diagram to include all the data and let the angle of the sector be θ.

$66 = 55\theta$ •

So $\quad \theta = 1.2$ radians

Area of plot $= \frac{1}{2}(55)^2(1.2)$ •

$\quad = 1815$ m^2

Draw a diagram to include all the data and let the angle of the sector be θ.

As the perimeter is given, first find length of arc AB.

Use the formula for arc length, $l = r\theta$.

Use the formula for area of a sector, $A = \frac{1}{2}r^2\theta$.

6.4 You can work out the area of a segment using radians.

■ The area of a segment in a circle of radius r is
  $A = \frac{1}{2}r^2 (\theta - \sin \theta)$

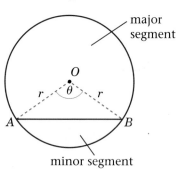

Example 10

Show that the area of the shaded segment in the circle shown is $\frac{1}{2}r^2 (\theta - \sin \theta)$

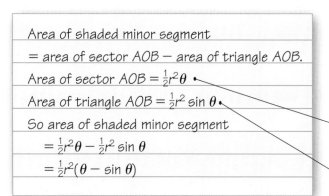

Area of shaded minor segment

$=$ area of sector AOB − area of triangle AOB.

Area of sector $AOB = \frac{1}{2}r^2\theta$ •

Area of triangle $AOB = \frac{1}{2}r^2 \sin \theta$ •

So area of shaded minor segment

$\quad = \frac{1}{2}r^2\theta - \frac{1}{2}r^2 \sin \theta$

$\quad = \frac{1}{2}r^2(\theta - \sin \theta)$

Use $A = \frac{1}{2}r^2\theta$.

Use $\frac{1}{2}ab \sin C$ from Section 2.7.

Example 11

In the diagram AB is the diameter of a circle of radius r cm, and $\angle BOC = \theta$ radians. Given that the area of $\triangle AOC$ is three times that of the shaded segment, show that $3\theta - 4\sin\theta = 0$.

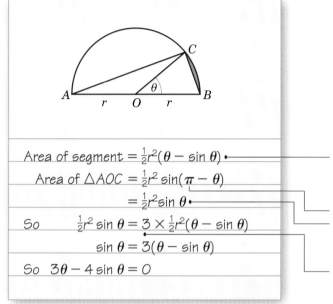

Area of segment $= \frac{1}{2}r^2(\theta - \sin\theta)$ •———— Area of segment area of sector − area of triangle.

Area of $\triangle AOC = \frac{1}{2}r^2\sin(\pi - \theta)$

$= \frac{1}{2}r^2\sin\theta$ •

$\angle AOB = \pi$ radians.

So $\frac{1}{2}r^2\sin\theta = 3 \times \frac{1}{2}r^2(\theta - \sin\theta)$

Remember from Section 2.3 that $\sin(180° - \theta°) = \sin\theta°$ so $\sin(\pi - \theta) = \sin\theta$.

$\sin\theta = 3(\theta - \sin\theta)$

So $3\theta - 4\sin\theta = 0$

Area of $\triangle AOC = 3 \times$ area of shaded segment.

(In Chapter 4 of Book C3 you will be able to find an approximation for θ.)

Exercise 6C

(*Note:* give non-exact answers to 3 significant figures.)

1 Find the area of the shaded sector in each of the following circles with centre C.
Leave your answer in terms of π, where appropriate.

a

8 cm
C
0.6^c

b

9 cm
C
$\frac{\pi}{6}$

c

C
$\frac{\pi}{5}$
1.2 cm

d

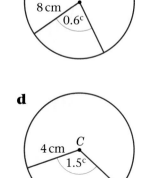

4 cm
C
1.5^c

e

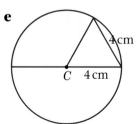

4 cm
C 4 cm

f

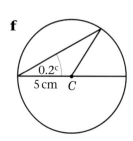

0.2^c
5 cm C

2 For the following circles with centre C, the area A of the shaded sector is given. Find the value of x in each case.

a

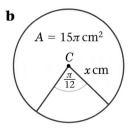

$A = 12\,\text{cm}^2$

b

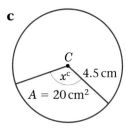

c

3 The arc AB of a circle, centre O and radius $6\,\text{cm}$, has length $4\,\text{cm}$.
Find the area of the minor sector AOB.

4 The chord AB of a circle, centre O and radius $10\,\text{cm}$, has length $18.65\,\text{cm}$ and subtends an angle of θ radians at O.

a Show that $\theta = 2.40$ (to 3 significant figures).

b Find the area of the minor sector AOB.

5 The area of a sector of a circle of radius $12\,\text{cm}$ is $100\,\text{cm}^2$.
Find the perimeter of the sector.

6 The arc AB of a circle, centre O and radius $r\,\text{cm}$, is such that $\angle AOB = 0.5$ radians.
Given that the perimeter of the minor sector AOB is $30\,\text{cm}$:

a Calculate the value of r.

b Show that the area of the minor sector AOB is $36\,\text{cm}^2$.

c Calculate the area of the segment enclosed by the chord AB and the minor arc AB.

7 In the diagram, AB is the diameter of a circle of radius $r\,\text{cm}$ and $\angle BOC = \theta$ radians. Given that the area of $\triangle COB$ is equal to that of the shaded segment, show that $\theta + 2\sin\theta = \pi$.

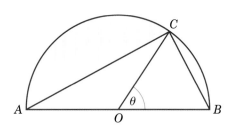

8 In the diagram, BC is the arc of a circle, centre O and radius $8\,\text{cm}$. The points A and D are such that $OA = OD = 5\,\text{cm}$. Given that $\angle BOC = 1.6$ radians, calculate the area of the shaded region.

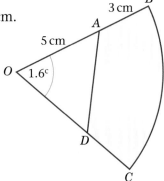

9 In the diagram, *AB* and *AC* are tangents to a circle, centre *O* and radius 3.6 cm. Calculate the area of the shaded region, given that $\angle BOC = \frac{2}{3}\pi$ radians.

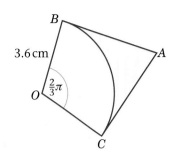

10 A chord *AB* subtends an angle of θ radians at the centre O of a circle of radius 6.5 cm. Find the area of the segment enclosed by the chord *AB* and the minor arc *AB*, when:

   a $\theta = 0.8$ **b** $\theta = \frac{2}{3}\pi$ **c** $\theta = \frac{4}{3}\pi$

11 An arc *AB* subtends an angle of 0.25 radians at the *circumference* of a circle, centre *O* and radius 6 cm. Calculate the area of the minor sector *OAB*.

12 In the diagram, AD and BC are arcs of circles with centre O, such that $OA = OD = r$ cm, $AB = DC = 8$ cm and $\angle BOC = \theta$ radians.

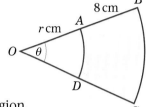

   a Given that the area of the shaded region is 48 cm², show that

$$r = \frac{6}{\theta} - 4.$$

   b Given also that $r = 10\theta$, calculate the perimeter of the shaded region.

13 A sector of a circle of radius 28 cm has perimeter *P* cm and area *A* cm². Given that $A = 4P$, find the value of *P*.

14 The diagram shows a triangular plot of land. The sides *AB*, *BC* and *CA* have lengths 12 m, 14 m and 10 m respectively. The lawn is a sector of a circle, centre *A* and radius 6 m.

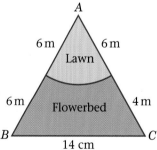

   a Show that $\angle BAC = 1.37$ radians, correct to 3 significant figures.

   b Calculate the area of the flowerbed.

Mixed exercise 6D

1 Triangle *ABC* is such that $AB = 5$ cm, $AC = 10$ cm and $\angle ABC = 90°$. An arc of a circle, centre *A* and radius 5 cm, cuts *AC* at *D*.

   a State, in radians, the value of $\angle BAC$.

   b Calculate the area of the region enclosed by *BC*, *DC* and the arc *BD*.

2 The diagram shows a minor sector *OMN* of a circle centre *O* and radius *r* cm. The perimeter of the sector is 100 cm and the area of the sector is *A* cm².

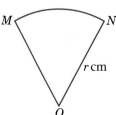

   a Show that $A = 50r - r^2$.

   b Given that *r* varies, find:

     i The value of *r* for which *A* is a maximum and show that *A* is a maximum.

     ii The value of $\angle MON$ for this maximum area.

     iii The maximum area of the sector *OMN*.

3 The diagram shows the triangle *OCD* with
$OC = OD = 17$ cm and $CD = 30$ cm. The mid-point
of *CD* is *M*. With centre *M*, a semicircular arc A_1 is
drawn on *CD* as diameter. With centre *O* and
radius 17 cm, a circular arc A_2 is drawn from *C* to
D. The shaded region *R* is bounded by the arcs A_1
and A_2. Calculate, giving answers to 2 decimal
places:

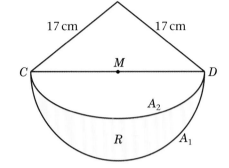

   a The area of the triangle *OCD*.

   b The angle *COD* in radians.

   c The area of the shaded region *R*.

      E

4 The diagram shows a circle, centre *O*, of radius 6 cm.
The points *A* and *B* are on the circumference of the circle.
The area of the shaded major sector is 80 cm².
Given that $\angle AOB = \theta$ radians, where $0 < \theta < \pi$,
calculate:

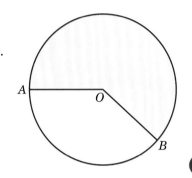

   a The value, to 3 decimal places, of θ.

   b The length in cm, to 2 decimal places, of the minor
arc *AB*.

      E

5 The diagram shows a sector *OAB* of a circle, centre *O* and
radius *r* cm. The length of the arc *AB* is *p* cm and $\angle AOB$ is
θ radians.

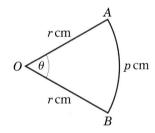

   a Find θ in terms of *p* and *r*.

   b Deduce that the area of the sector is $\frac{1}{2}pr$ cm².

Given that $r = 4.7$ and $p = 5.3$, where each has been measured to
1 decimal place, find, giving your answer to 3 decimal places:

   c The least possible value of the area of the sector.

   d The range of possible values of θ.

      E

6 The diagram shows a circle centre *O* and radius 5 cm.
The length of the minor arc *AB* is 6.4 cm.

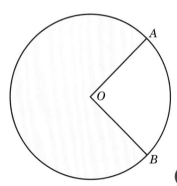

   a Calculate, in radians, the size of the acute angle *AOB*.

The area of the minor sector *AOB* is R_1 cm² and the area
of the shaded major sector *AOB* is R_2 cm².

   b Calculate the value of R_1.

   c Calculate $R_1 : R_2$ in the form $1 : p$, giving the value of *p*
to 3 significant figures.

      E

7

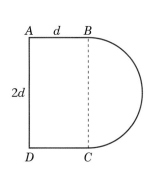

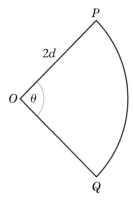

Shape X Shape Y

The diagrams show the cross-sections of two drawer handles.

Shape X is a rectangle $ABCD$ joined to a semicircle with BC as diameter. The length $AB = d$ cm and $BC = 2d$ cm. Shape Y is a sector OPQ of a circle with centre O and radius $2d$ cm. Angle POQ is θ radians.

Given that the areas of shapes X and Y are equal:

a Prove that $\theta = 1 + \frac{1}{4}\pi$.

Using this value of θ, and given that $d = 3$, find in terms of π:

b The perimeter of shape X.

c The perimeter of shape Y.

d Hence find the difference, in mm, between the perimeters of shapes X and Y.

E

8 The diagram shows a circle with centre O and radius 6 cm. The chord PQ divides the circle into a minor segment R_1 of area A_1 cm² and a major segment R_2 of area A_2 cm². The chord PQ subtends an angle θ radians at O.

a Show that $A_1 = 18(\theta - \sin \theta)$.

Given that $A_2 = 3A_1$ and $f(\theta) = 2\theta - 2 \sin \theta - \pi$:

b Prove that $f(\theta) = 0$.

c Evaluate $f(2.3)$ and $f(2.32)$ and deduce that $2.3 < \theta < 2.32$.

E

9 Triangle ABC has $AB = 9$ cm, $BC = 10$ cm and $CA = 5$ cm. A circle, centre A and radius 3 cm, intersects AB and AC at P and Q respectively, as shown in the diagram.

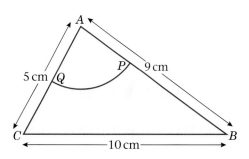

a Show that, to 3 decimal places,
$\angle BAC = 1.504$ radians.

b Calculate:
   i The area, in cm², of the sector APQ.
   ii The area, in cm², of the shaded region $BPQC$.
   iii The perimeter, in cm, of the shaded region $BPQC$.

E

10 The diagram shows the sector OAB of a circle of radius r cm. The area of the sector is 15 cm² and $\angle AOB = 1.5$ radians.

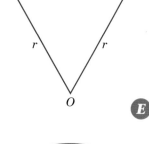

  a Prove that $r = 2\sqrt{5}$.

  b Find, in cm, the perimeter of the sector OAB.

  The segment R, shaded in the diagram, is enclosed by the arc AB and the straight line AB.

  c Calculate, to 3 decimal places, the area of R.

  (E)

11 The shape of a badge is a sector ABC of a circle with centre A and radius AB, as shown in the diagram. The triangle ABC is equilateral and has perpendicular height 3 cm.

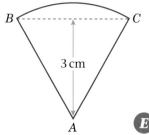

  a Find, in surd form, the length of AB.

  b Find, in terms of π, the area of the badge.

  c Prove that the perimeter of the badge is $\dfrac{2\sqrt{3}}{3}(\pi + 6)$ cm.

  (E)

12 There is a straight path of length 70 m from the point A to the point B. The points are joined also by a railway track in the form of an arc of the circle whose centre is C and whose radius is 44 m, as shown in the diagram.

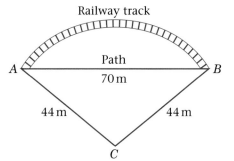

  a Show that the size, to 2 decimal places, of $\angle ACB$ is 1.84 radians.

  b Calculate:

   i The length of the railway track.
   ii The shortest distance from C to the path.
   iii The area of the region bounded by the railway track and the path.

  (E)

13 The diagram shows the cross-section $ABCD$ of a glass prism. $AD = BC = 4$ cm and both are at right angles to DC. AB is the arc of a circle, centre O and radius 6 cm. Given that $\angle AOB = 2\theta$ radians, and that the perimeter of the cross-section is $2(7 + \pi)$ cm:

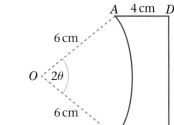

  a Show that $(2\theta + 2\sin\theta - 1) = \dfrac{\pi}{3}$.

  b Verify that $\theta = \dfrac{\pi}{6}$.

  c Find the area of the cross-section.

14 Two circles C_1 and C_2, both of radius 12 cm, have centres O_1 and O_2 respectively. O_1 lies on the circumference of C_2; O_2 lies on the circumference of C_1. The circles intersect at A and B, and enclose the region R.

  a Show that $\angle AO_1B = \frac{2}{3}\pi$ radians.

  b Hence write down, in terms of π, the perimeter of R.

  c Find the area of R, giving your answer to 3 significant figures.

Summary of key points

1   If the arc AB has length r, then $\angle AOB$ is 1 radian
(1^c or 1 rad).

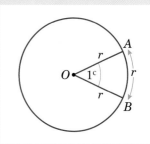

2   A radian is the angle subtended at the centre of a circle
by an arc whose length is equal to that of the radius of
the circle.

3   1 radian $= \dfrac{180°}{\pi}$.

4   The length of an arc of a circle is $l = r\theta$.

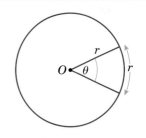

5   The area of a sector is $A = \frac{1}{2}r^2\theta$.

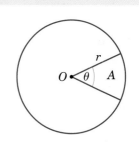

6   The area of a segment in a circle is $A = \frac{1}{2}r^2(\theta - \sin\theta)$.

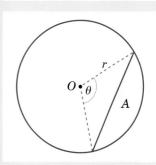

7 Geometric sequences and series

This chapter shows you how to calculate terms in geometric sequences, and how to find the sum of both finite and infinite geometric series.

7.1 The following sequences are called geometric sequences. To get from one term to the next we multiply by the same number each time. This number is called the common ratio, r.

1, 2, 4, 8, 16, ...
100, 25, 6.25, 1.5625, ...
2, −6, 18, −54, 162, ...

Example 1

Find the common ratios in the following geometric sequences:

a 2, 10, 50, 250, ...

b 90, −30, 10, $-3\frac{1}{3}$

a 2, 10, 50, 250, ...	Use u_1, u_2 etc. to refer to the individual terms in a sequence. Here $u_1 = 2$, $u_2 = 10$, $u_3 = 50$.
Common ratio $= \dfrac{10}{2} = 5$	To find the common ratio calculate $\dfrac{u_2}{u_1}$ or $\dfrac{u_3}{u_2}$.

a 90, −30, 10, $-3\frac{1}{3}$	Common ratio $= \dfrac{u_2}{u_1}$.
Common ratio $= \dfrac{-30}{90} = -\dfrac{1}{3}$	A common ratio can be negative or a fraction (or both).

Exercise 7A

1 Which of the following are geometric sequences? For the ones that are, give the value of 'r' in the sequence:

a 1, 2, 4, 8, 16, 32, ... **b** 2, 5, 8, 11, 14, ...

c 40, 36, 32, 28, ... **d** 2, 6, 18, 54, 162, ...

e 10, 5, 2.5, 1.25, ... **f** 5, −5, 5, −5, 5, ...

g 3, 3, 3, 3, 3, 3, 3, ... **h** 4, −1, 0.25, −0.0625, ...

2 Continue the following geometric sequences for three more terms:

    a 5, 15, 45, ...   **b** 4, −8, 16, ...   **c** 60, 30, 15, ...

    d $1, \frac{1}{4}, \frac{1}{16}$, ...   **e** $1, p, p^2$, ...   **f** $x, -2x^2, 4x^3$, ...

3 If 3, x and 9 are the first three terms of a geometric sequence. Find:

    a The exact value of x.

    b The exact value of the 4th term.

> **Hint for question 3:**
> In a geometric sequence the common ratio can be calculated by $\frac{u_2}{u_1}$ or $\frac{u_3}{u_2}$.

7.2 You can define a geometric sequence using the first term a and the common ratio r:

$$a, \qquad ar, \qquad ar^2, \qquad ar^3, \ldots \quad ar^{n-1}$$
$$\uparrow \qquad\quad \uparrow \qquad\quad \uparrow \qquad\quad \uparrow \qquad\quad\quad \uparrow$$
1st term 2nd term 3rd term 4th term nth term

> Sometimes a geometric sequence is called a geometric progression.

> **Hint:** Look at the relationship between the position of the term in the sequence and the index of the term. You should be able to see that the index of r is one less than its position in the sequence. So the nth term of a geometric sequence is ar^{n-1}.

Example 2

Find the **i** 10th and **ii** nth terms
in the following geometric sequences:

    a 3, 6, 12, 24, ...   **b** 40, −20, 10, −5, ...

---

a 3, 6, 12, 24, ...

    i 10th term = $3 \times (2)^9$

              = 3×512

              = 1536

    ii nth term = $3 \times 2^{n-1}$

For this sequence $a = 3$ and $r = \frac{6}{3} = 2$.

For the 10th term use ar^{n-1} with $a = 3$, $r = 2$ and $n = 10$.

For the nth term use ar^{n-1} with $a = 3$ and $r = 2$.

---

b 40, −20, 10, −5, ...

    i 10th term = $40 \times (-\frac{1}{2})^9$

              = $40 \times -\frac{1}{512}$

              = $-\frac{5}{64}$

    ii nth term = $40 \times (-\frac{1}{2})^{n-1}$

              = $5 \times 8 \times (-\frac{1}{2})^{n-1}$

              = $5 \times 2^3 \times (-\frac{1}{2})^{n-1}$

              = $(-1)^{n-1} \times \frac{5}{2^{n-4}}$

For this sequence $a = 40$ and $r = -\frac{20}{40} = -\frac{1}{2}$.

Use ar^{n-1} with $a = 40$, $r = -\frac{1}{2}$ and $n = 10$.

Use ar^{n-1} with $a = 40$, $r = -\frac{1}{2}$ and $n = n$.

Use laws of indices $\frac{x^m}{x^n} = \frac{1}{x^{n-m}}$.

So $2^3 \times \dfrac{1}{2^{n-1}} = \dfrac{1}{2^{n-1-3}}$.

Example 3

The second term of a geometric sequence is 4 and the 4th term is 8.
Find the exact values of **a** the common ratio, **b** first term and **c** the
10th term:

a 2nd term = 4, $ar = 4$ ① Using nth term = ar^{n-1}
with $n = 2$
 4th term = 6, $ar^3 = 8$ ② and $n = 4$.

 ② ÷ ① $r^2 = 2$

 $r = \sqrt{2}$

So Common ratio = $\sqrt{2}$

b Substitute back in ① $a\sqrt{2} = 4$ Divide equation by $\sqrt{2}$.

 $a = \dfrac{4}{\sqrt{2}}$

 $= \dfrac{4\sqrt{2}}{2}$ To rationalise $\dfrac{4}{\sqrt{2}}$, multiply top and bottom

 $a = 2\sqrt{2}$ by $\sqrt{2}$.

So first term = $2\sqrt{2}$

c 10th term = ar^9

 $= 2\sqrt{2}(\sqrt{2})^9$ Substitute the values of $a(= 2\sqrt{2})$ and
$r(= \sqrt{2})$ back into ar^{n-1} with $n = 10$.
 $= 2(\sqrt{2})^{10}$

 $= 2 \times 2^5$ $(\sqrt{2})^{10} = (2^{\frac{1}{2}})^{10} = 2^{\frac{1}{2} \times 10} = 2^5$

 $= 2^6$

 $= 64$

So 10th term = 64

Example 4

The numbers 3, x and $(x + 6)$ form the first three terms of a positive geometric sequence. Find:

a The possible values of x.   **b** The 10th term of the sequence.

a

$$\frac{u_2}{u_1} = \frac{u_3}{u_2}$$ The sequence is geometric so $\frac{u_2}{u_1} = \frac{u_3}{u_2}$.

$$\frac{x}{3} = \frac{x + 6}{x}$$ Cross multiply.

$$x^2 = 3(x + 6)$$
$$x^2 = 3x + 18$$
$$x^2 - 3x - 18 = 0$$
$$(x - 6)(x + 3) = 0$$ Factorise.
$$x = 6 \text{ or } -3$$

So x is either 6 or -3, but there are no negative terms so $x = 6$.   If there are no negative terms then -3 cannot be an answer.

Accept $x = 6$, as terms are positive.

b 10th term $= ar^9$

$= 3 \times 2^9$ Use the formula nth term $= ar^{n-1}$ with $n = 9$,

$= 3 \times 512$ $a = 3$ and $r = \frac{x}{3} = \frac{6}{3} = 2$.

$= 1536$

The 10th term is 1536.

Exercise 7B

1 Find the sixth, tenth and nth terms of the following geometric sequences:

   a 2, 6, 18, 54, ...   **b** 100, 50, 25, 12.5, ...
   c 1, -2, 4, -8, ...   **d** 1, 1.1, 1.21, 1.331, ...

2 The nth term of a geometric sequence is $2 \times (5)^n$. Find the first and 5th terms.

3 The sixth term of a geometric sequence is 32 and the 3rd term is 4. Find the first term and the common ratio.

4 Given that the first term of a geometric sequence is 4, and the third is 1, find possible values for the 6th term.

5 The expressions $x - 6$, $2x$ and x^2 form the first three terms of a geometric progression. By calculating two different expressions for the common ratio, form and solve an equation in x to find possible values of the first term.

7.3 You can use geometric sequences to solve problems involving growth and decay, e.g. interest rates, population growth and decline.

Example **5**

Andy invests £A at a rate of interest 4% per annum.
After 5 years it will be worth £10 000.
How much (to the nearest penny) will it be worth after 10 years?

After 1 year it will be worth £A × 1.04

After 2 years it will be worth

£A × 1.04 × 1.04 = £A × 1.04^2

So after 5 years investment is worth

£A × 1.04^5 ────── This is a geometric sequence where a = £A and r = 1.04.

A × 1.04^5 = £10 000 ────── After 5 years the investment is worth £10 000.

$A = \dfrac{£10\,000}{1.04^5}$ ────── Divide by 1.04^5.

= £8219.27

The initial investment A = £8219.27

After 10 years the investment is worth

A × 1.04^{10} ────── Use the exact value of A.

$A \times r^{10} = \dfrac{10\,000}{1.04^5} \times 1.04^{10}$

= 10 000 × 1.04^5 ────── Use laws of indices $\dfrac{x^m}{x^n} = x^{m-n}$.

= 12 166.529 02

= £12 166.53 ────── Put to the nearest penny.

If property values are increasing at 4% per annum, the multiplication factor is 1.04 (100% + 4%). So you multiply by 1.04 for each year you have this rise. However, if unemployment is coming down by 4% per annum, then the factor is 0.96 (100% − 4%).

Example 6

What is the first term in the geometric progression 3, 6, 12, 24, ... to exceed 1 million?

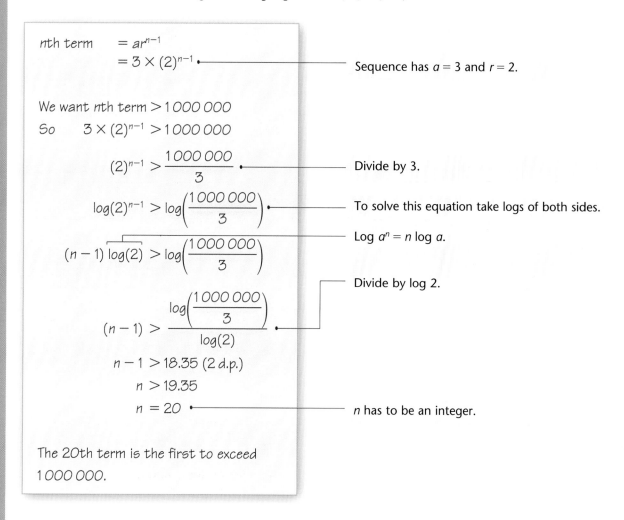

nth term $= ar^{n-1}$
$= 3 \times (2)^{n-1}$ ———————— Sequence has $a = 3$ and $r = 2$.

We want nth term $> 1\,000\,000$

So $3 \times (2)^{n-1} > 1\,000\,000$

$(2)^{n-1} > \dfrac{1\,000\,000}{3}$ ———————— Divide by 3.

$\log(2)^{n-1} > \log\left(\dfrac{1\,000\,000}{3}\right)$ ———————— To solve this equation take logs of both sides.

$(n-1)\log(2) > \log\left(\dfrac{1\,000\,000}{3}\right)$ ———————— Log $a^n = n \log a$.

$(n-1) > \dfrac{\log\left(\dfrac{1\,000\,000}{3}\right)}{\log(2)}$ ———————— Divide by log 2.

$n - 1 > 18.35$ (2 d.p.)

$n > 19.35$

$n = 20$ ———————— n has to be an integer.

The 20th term is the first to exceed 1 000 000.

Exercise 7C

1 A population of ants is growing at a rate of 10% a year. If there were 200 ants in the initial population, write down the number after

    a 1 year,   **b** 2 years,   **c** 3 years and   **d** 10 years.

2 A motorcycle has four gears. The maximum speed in bottom gear is $40 \, \text{km h}^{-1}$ and the maximum speed in top gear is $120 \, \text{km h}^{-1}$. Given that the maximum speeds in each successive gear form a geometric progression, calculate, in km h^{-1} to one decimal place, the maximum speeds in the two intermediate gears.

3 A car depreciates in value by 15% a year. If it is worth £11 054.25 after 3 years, what was its new price and when will it first be worth less than £5000?

4 The population decline in a school of whales can be modelled by a geometric progression. Initially there were 80 whales in the school. Four years later there were 40. Find out how many there will be at the end of the fifth year. (Round to the nearest whole number.)

5 Find which term in the progression 3, 12, 48, ... is the first to exceed 1 000 000.

6 A virus is spreading such that the number of people infected increases by 4% a day. Initially 100 people were diagnosed with the virus. How many days will it be before 1000 are infected?

7 I invest £A in the bank at a rate of interest of 3.5% per annum. How long will it be before I double my money?

8 The fish in a particular area of the North Sea are being reduced by 6% each year due to overfishing. How long would it be before the fish stocks are halved?

7.4 You need to be able to find the sum of a geometric series.

Example 7

Find the general term for the sum of the first n terms of a geometric series a, ar, ar^2, ... , ar^n.

Let $S_n = a + ar + ar^2 + ar^3 + ... + ar^{n-2} + ar^{n-1}$ ①

$rS_n = ar + ar^2 + ar^3 + ... + ar^{n-1} + ar^n$ ② ———— Multiply by r.

① − ② gives $S_n - rS_n = a - ar^n$ ———— Subtract rS_n from S_n.

$S_n(1 - r) = a(1 - r^n)$ ———— Take out the common factor.

$S_n = \dfrac{a(1 - r^n)}{1 - r}$ ———— Divide by $(1 - r)$.

■ **The general rule for the sum of a geometric series is** $S_n = \dfrac{a(r^n - 1)}{r - 1}$ **or** $\dfrac{a(1 - r^n)}{1 - r}$

Example 8

Find the sum of the following series:

a $2 + 6 + 18 + 54 + ...$ (for 10 terms)

b $1024 - 512 + 256 - 128 + ... + 1$

a Series is

$2 + 6 + 18 + 54 + ...$ (for 10 terms) ———— As in all questions, write down what is given.

So $a = 2$, $r = \frac{6}{2} = 3$ and $n = 10$

So $S_{10} = \dfrac{2(3^{10} - 1)}{3 - 1} = 59\,048$ ———— As $r = 3$ (>1), use the formula $S_n = \dfrac{a(r^n - 1)}{r - 1}$.

b Series is

$$1024 - 512 + 256 - 128 + \ldots + 1$$

So $a = 1024$, $r = -\frac{512}{1024} = -\frac{1}{2}$

and nth term $= 1$

$$1024\left(-\frac{1}{2}\right)^{n-1} = 1$$ — First solve $ar^{n-1} = 1$ to find n.

$$(-2)^{n-1} = 1024$$

$$2^{n-1} = 1024$$ — $(-2)^{n-1} = (-1)^{n-1}(2^{n-1}) = 1024$, so $(-1)^{n-1}$ must be positive and $2^{n-1} = 1024$.

$$n - 1 = \frac{\log 1024}{\log 2}$$

$$n - 1 = 10$$ — $1024 = 2^{10}$

$$n = 11$$

So $S_{11} = \dfrac{1024\left[1 - \left(-\frac{1}{2}\right)^{11}\right]}{1 - \left(-\frac{1}{2}\right)}$ — As $r = -\frac{1}{2}$ (<1) we use the formula $S_n = \dfrac{a(1 - r^n)}{1 - r}$.

$$= \frac{1024\left(1 + \frac{1}{2048}\right)}{1 + \frac{1}{2}}$$

$$= \frac{1024.5}{\frac{3}{2}} = 683$$

Example 9

An investor invests £2000 on January 1st every year in a savings account that guarantees him 4% per annum for life. If interest is calculated on the 31st of December each year, how much will be in the account at the end of the 10th year?

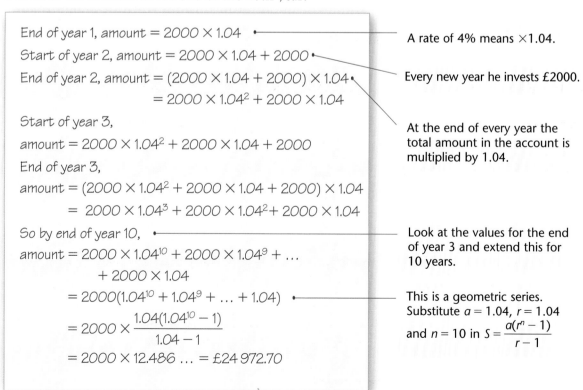

End of year 1, amount $= 2000 \times 1.04$ — A rate of 4% means $\times 1.04$.

Start of year 2, amount $= 2000 \times 1.04 + 2000$ — Every new year he invests £2000.

End of year 2, amount $= (2000 \times 1.04 + 2000) \times 1.04$
$$= 2000 \times 1.04^2 + 2000 \times 1.04$$

Start of year 3,
amount $= 2000 \times 1.04^2 + 2000 \times 1.04 + 2000$

End of year 3,
amount $= (2000 \times 1.04^2 + 2000 \times 1.04 + 2000) \times 1.04$
$$= 2000 \times 1.04^3 + 2000 \times 1.04^2 + 2000 \times 1.04$$

— At the end of every year the total amount in the account is multiplied by 1.04.

So by end of year 10,
amount $= 2000 \times 1.04^{10} + 2000 \times 1.04^9 + \ldots$
$$+ 2000 \times 1.04$$

— Look at the values for the end of year 3 and extend this for 10 years.

$$= 2000(1.04^{10} + 1.04^9 + \ldots + 1.04)$$

— This is a geometric series. Substitute $a = 1.04$, $r = 1.04$ and $n = 10$ in $S = \dfrac{a(r^n - 1)}{r - 1}$

$$= 2000 \times \frac{1.04(1.04^{10} - 1)}{1.04 - 1}$$

$$= 2000 \times 12.486\ldots = £24\,972.70$$

Example 10

Find the least value of n such that the sum of $1 + 2 + 4 + 8 + \ldots$ to n terms would exceed $2\,000\,000$.

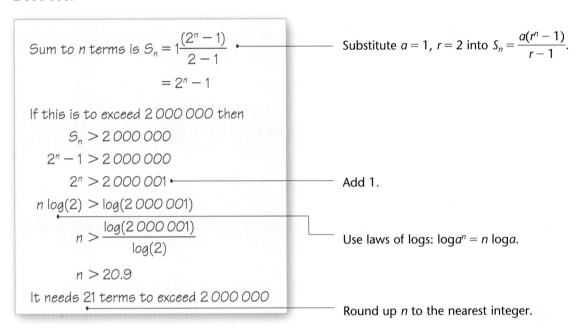

Sum to n terms is $S_n = 1\dfrac{(2^n - 1)}{2 - 1}$ ——— Substitute $a = 1$, $r = 2$ into $S_n = \dfrac{a(r^n - 1)}{r - 1}$.

$$= 2^n - 1$$

If this is to exceed $2\,000\,000$ then

$$S_n > 2\,000\,000$$

$$2^n - 1 > 2\,000\,000$$

$$2^n > 2\,000\,001$$ ——— Add 1.

$$n \log(2) > \log(2\,000\,001)$$

$$n > \frac{\log(2\,000\,001)}{\log(2)}$$ ——— Use laws of logs: $\log a^n = n \log a$.

$$n > 20.9$$

It needs 21 terms to exceed $2\,000\,000$ ——— Round up n to the nearest integer.

Example 11

Find $\displaystyle\sum_{r=1}^{10} (3 \times 2^r)$.

$$S_{10} = \sum_{r=1}^{10} (3 \times 2^r)$$ ——— 'Σ' means 'sum of' – in this case the sum of (3×2^r) from $r = 1$ to $r = 10$.

$$= 3 \times 2^1 + 3 \times 2^2 + 3 \times 2^3$$
$$+ \ldots + 3 \times 2^{10}$$

$$= 3(2^1 + 2^2 + 2^3 + \ldots + 2^{10})$$ ——— This is a geometric series with $a = 2$, $r = 2$ and $n = 10$.

$$= 3 \times 2\frac{(2^{10} - 1)}{2 - 1}$$ ——— Use $s = \dfrac{a(r^n - 1)}{r - 1}$

So $S_{10} = 6138$

Exercise 7D

1 Find the sum of the following geometric series (to 3 d.p. if necessary):

a $1 + 2 + 4 + 8 + \ldots$ (8 terms)   **b** $32 + 16 + 8 + \ldots$ (10 terms)

c $4 - 12 + 36 - 108 \ldots$ (6 terms)   **d** $729 - 243 + 81 - \ldots -\frac{1}{3}$

e $\displaystyle\sum_{r=1}^{6} 4^r$

f $\displaystyle\sum_{r=1}^{8} 2 \times (3)^r$

g $\displaystyle\sum_{r=1}^{10} 6 \times (\tfrac{1}{2})^r$

h $\displaystyle\sum_{r=0}^{5} 60 \times (-\tfrac{1}{3})^r$

2 The sum of the first three terms of a geometric series is 30.5. If the first term is 8, find possible values of *r*.

3 The man who invented the game of chess was asked to name his reward. He asked for 1 grain of corn to be placed on the first square of his chessboard, 2 on the second, 4 on the third and so on until all 64 squares were covered. He then said he would like as many grains of corn as the chessboard carried. How many grains of corn did he claim as his prize?

4 Jane invests £4000 at the start of every year. She negotiates a rate of interest of 4% per annum, which is paid at the end of the year. How much is her investment worth at the end of **a** the 10th year and **b** the 20th year?

5 A ball is dropped from a height of 10 m. It bounces to a height of 7 m and continues to bounce. Subsequent heights to which it bounces follow a geometric sequence. Find out:

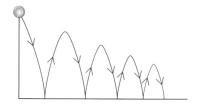

   a How high it will bounce after the fourth bounce.

   b The total distance travelled after it hits the ground for the sixth time.

6 Find the least value of *n* such that the sum $3 + 6 + 12 + 24 + \ldots$ to *n* terms would first exceed 1.5 million.

7 Find the least value of *n* such that the sum $5 + 4.5 + 4.05 + \ldots$ to *n* terms would first exceed 45.

8 Richard is sponsored to cycle 1000 miles over a number of days. He cycles 10 miles on day 1, and increases this distance by 10% a day. How long will it take him to complete the challenge? What was the greatest number of miles he completed in a single day?

9 A savings scheme is offering a rate of interest of 3.5% per annum for the lifetime of the plan. Alan wants to save up £20 000. He works out that he can afford to save £500 every year, which he will deposit on January 1st. If interest is paid on 31st of December, how many years will it be before he has saved up his £20 000?

7.5 **You need to be able to find the sum to infinity of a convergent geometric series.**

Consider the series $S = 3 + 1.5 + 0.75 + 0.375 + \ldots$

No matter how many terms of the series you take, the sum never exceeds a certain number. We call this number the limit of the sum, or more often, its sum to infinity.

We can find out what this limit is.

As $a = 3$ and $r = \frac{1}{2}$, $S = \dfrac{a(1 - r^n)}{1 - r} = \dfrac{3(1 - (\frac{1}{2})^n)}{1 - \frac{1}{2}} = 6(1 - (\frac{1}{2})^n)$

If we replace *n* with certain values to find the sum we find that

   when $n = 3$, $S_3 = 5.25$
   when $n = 5$, $S_5 = 5.8125$
   when $n = 10$, $S_{10} = 5.9994$
   when $n = 20$, $S_{20} = 5.999\,994$

You can see that as n gets larger, S becomes closer and closer to 6.

We say that this infinite series is **convergent**, and has a sum to infinity of 6. Convergent means the series tends towards a specific value as more terms are added.

Not all series converge. The reason that this one does is that the terms of the sequence are getting smaller.

This happens because $-1 < r < 1$.

The sum to infinity of a series exists only if $-1 < r < 1$.

$$S_n = \frac{a(1 - r^n)}{1 - r}$$

If $-1 < r < 1$, $r^n \to 0$ as $n \to \infty$

$$S_\infty = \frac{a(1 - 0)}{1 - r} = \frac{a}{1 - r}$$

Hint: You can write 'the sum to infinity' is S_∞.

■ The sum to infinity of a geometric series is $\dfrac{a}{1 - r}$
   if $|r| < 1$.

Hint: $|r|$ means $-1 < r < 1$.

Example 12

Find the sums to infinity of the following series:

a $40 + 10 + 2.5 + 0.625 + \ldots$

b $1 + \dfrac{1}{p} + \dfrac{1}{p^2} + \ldots$

a $40 + 10 + 2.5 + 0.625 + \ldots$

In this series $a = 40$ and $r = \dfrac{10}{40} = \dfrac{1}{4}$

$-1 < r < 1$, so S_∞ exists

$$S = \frac{a}{1 - r} = \frac{40}{1 - \frac{1}{4}} = \frac{40}{\frac{3}{4}} = \frac{160}{3}$$

Always write down the values of a and r, using $\dfrac{u_2}{u_1}$ for r.

Substitute $a = 40$ and $r = \dfrac{10}{40} = \dfrac{1}{4}$ into $S = \dfrac{a}{1 - r}$.

b $1 + \dfrac{1}{p} + \dfrac{1}{p^2} + \ldots$

In this series $a = 1$ and $r = \dfrac{u_2}{u_1} = \dfrac{\frac{1}{p}}{1} = \dfrac{1}{p}$

S will exist if $\left|\dfrac{1}{p}\right| < 1$ so $p > 1$.

If $p > 1$, $S_\infty = \dfrac{1}{1 - \frac{1}{p}}$

$$= \frac{p}{p - 1}$$

Multiply top and bottom by p.

Example 13

The sum to 4 terms of a geometric series is 15 and the sum to infinity is 16.

a Find the possible values of r.

b Given that the terms are all positive, find the first term in the series.

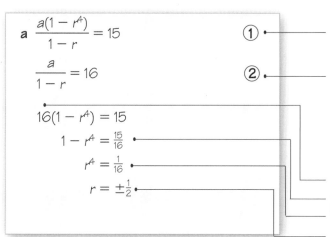

a $\dfrac{a(1 - r^4)}{1 - r} = 15$ ① •————— $S_4 = 15$ so use the formula $S_n = \dfrac{a(1 - r^n)}{1 - r}$ with $n = 4$.

$\dfrac{a}{1 - r} = 16$ ② •————— $S_\infty = 16$ so use the formula $S_\infty = \dfrac{a}{1 - r}$ with $S_\infty = 16$.

$16(1 - r^4) = 15$

$1 - r^4 = \dfrac{15}{16}$ •———   Solve equations simultaneously.

$r^4 = \dfrac{1}{16}$ •———   Replace $\dfrac{a}{1 - r}$ by 16 in equation ①

$r = \pm\dfrac{1}{2}$ •———   Divide by 16.

     Rearrange.

     Take the 4th root of $\dfrac{1}{16}$.

b As all terms positive, $r = +\dfrac{1}{2}$

Substitute $r = +\dfrac{1}{2}$ back into equation ② to find a

$\dfrac{a}{1 - \frac{1}{2}} = 16$

$16\left(1 - \dfrac{1}{2}\right) = a$

$a = 8$

The first term in the series is 8.

Exercise 7E

1 Find the sum to infinity, if it exists, of the following series:

    a $1 + 0.1 + 0.01 + 0.001 + \dots$ **b** $1 + 2 + 4 + 8 + 16 + \dots$

    c $10 - 5 + 2.5 - 1.25 + \dots$ **d** $2 + 6 + 10 + 14$

    e $1 + 1 + 1 + 1 + 1 + \dots$ **f** $3 + 1 + \frac{1}{3} + \frac{1}{9} + \dots$

    g $0.4 + 0.8 + 1.2 + 1.6 + \dots$ **h** $9 + 8.1 + 7.29 + 6.561 + \dots$

    i $1 + r + r^2 + r^3 + \dots$ **j** $1 - 2x + 4x^2 - 8x^3 + \dots$

2 Find the common ratio of a geometric series with a first term of 10 and a sum to infinity of 30.

3 Find the common ratio of a geometric series with a first term of -5 and a sum to infinity of -3.

4 Find the first term of a geometric series with a common ratio of $\frac{2}{3}$ and a sum to infinity of 60.

5 Find the first term of a geometric series with a common ratio of $-\frac{1}{3}$ and a sum to infinity of 10.

6 Find the fraction equal to the recurring decimal 0.232 323 232 3.

7 Find $\displaystyle\sum_{r=1}^{\infty} 4(0.5)^r$.

> **Hint for question 6:** Write
> 0.232 323 232 3 as
> $\frac{23}{100} + \frac{23}{10\,000} + \frac{232}{1\,000\,000} + \dots$

8 A ball is dropped from a height of 10 m. It bounces to a height of 6 m, then 3.6, and so on following a geometric sequence.
Find the total distance travelled by the ball.

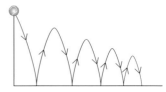

9 The sum to three terms of geometric series is 9 and its sum to infinity is 8. What could you deduce about the common ratio. Why? Find the first term and common ratio.

10 The sum to infinity of a geometric series is three times the sum to 2 terms. Find all possible values of the common ratio.

Mixed exercise 7F

1 State which of the following series are geometric. For the ones that are, give the value of the common ratio r.

a $4 + 7 + 10 + 13 + 16 + \dots$ **b** $4 + 6 + 9 + 13.5 + \dots$

c $20 + 10 + 5 + 2.5 + \dots$ **d** $4 - 8 + 16 - 32 + \dots$

e $4 - 2 - 8 - 14 - \dots$ **f** $1 + 1 + 1 + 1 + \dots$

2 Find the 8th and nth terms of the following geometric sequences:

a 10, 7, 4.9, ...   **b** 5, 10, 20, ...

c 4, −4, 4, ...   **d** 3, −1.5, 0.75, ...

3 Find the sum to 10 terms and of the following geometric series:

a $4 + 8 + 16 + \dots$ **b** $30 - 15 + 7.5, \dots$

c $5 + 5 + 5, \dots$ **d** $2 + 0.8 + 0.32, \dots$

4 Determine which of the following geometric series converge. For the ones that do, give the limiting value of this sum (i.e. S_∞).

a $6 + 2 + \frac{2}{3} + \dots$ **b** $4 - 2 + 1 - \dots$

c $5 + 10 + 20 + \dots$ **d** $4 + 1 + 0.25 + \dots$

5 A geometric series has third term 27 and sixth term 8:

a Show that the common ratio of the series is $\frac{2}{3}$.

b Find the first term of the series.

c Find the sum to infinity of the series.

d Find, to 3 significant figures, the difference between the sum of the first 10 terms of the series and the sum to infinity of the series.

E

6 The second term of a geometric series is 80 and the fifth term of the series is 5.12:

  a Show that the common ratio of the series is 0.4.

  Calculate:

  b The first term of the series.

  c The sum to infinity of the series, giving your answer as an exact fraction.

  d The difference between the sum to infinity of the series and the sum of the first 14 terms of the series, giving your answer in the form $a \times 10^n$, where $1 \leqslant a < 10$ and n is an integer.

7 The nth term of a sequence is u_n, where $u_n = 95(\frac{4}{5})^n$, $n = 1, 2, 3, \ldots$

  a Find the value of u_1 and u_2.

  Giving your answers to 3 significant figures, calculate:

  b The value of u_{21}.

  c $\sum_{n=1}^{15} u_n$

  d Find the sum to infinity of the series whose first term is u_1 and whose nth term is u_n.

8 A sequence of numbers $u_1, u_2, \ldots, u_n, \ldots$ is given by the formula $u_n = 3(\frac{2}{3})^n - 1$ where n is a positive integer.

  a Find the values of u_1, u_2 and u_3.

  b Show that $\sum_{n=1}^{15} u_n = -9.014$ to 4 significant figures.

  c Prove that $u_{n+1} = 2(\frac{2}{3})^n - 1$.

9 The third and fourth terms of a geometric series are 6.4 and 5.12 respectively. Find:

  a The common ratio of the series.

  b The first term of the series.

  c The sum to infinity of the series.

  d Calculate the difference between the sum to infinity of the series and the sum of the first 25 terms of the series.

10 The price of a car depreciates by 15% per annum. If its new price is £20 000, find:

  a A formula linking its value £V with its age a years.

  b Its value after 5 years.

  c The year in which it will be worth less than £4000.

11 The first three terms of a geometric series are $p(3q + 1)$, $p(2q + 2)$ and $p(2q - 1)$ respectively, where p and q are non-zero constants.

  a Use algebra to show that one possible value of q is 5 and to find the other possible value of q.

  b For each possible value of q, calculate the value of the common ratio of the series.

  Given that $q = 5$ and that the sum to infinity of the geometric series is 896, calculate:

  c The value of p.

  d The sum, to 2 decimal places, of the first twelve terms of the series.

12 A savings scheme pays 5% per annum compound interest. A deposit of £100 is invested in this scheme at the start of each year.

    a Show that at the start of the third year, after the annual deposit has been made, the amount in the scheme is £315.25.

    b Find the amount in the scheme at the start of the fortieth year, after the annual deposit has been made.

13 A competitor is running in a 25 km race. For the first 15 km, she runs at a steady rate of $12 \, \text{km h}^{-1}$. After completing 15 km, she slows down and it is now observed that she takes 20% longer to complete each kilometre than she took to complete the previous kilometre.

    a Find the time, in hours and minutes, the competitor takes to complete the first 16 km of the race.

    The time taken to complete the rth kilometre is u_r hours.

    b Show that, for $16 \leq r \leq 25$, $u_r = \frac{1}{12}(1.2)^{r-15}$.

    c Using the answer to **b**, or otherwise, find the time, to the nearest minute, that she takes to complete the race.

14 A liquid is kept in a barrel. At the start of a year the barrel is filled with 160 litres of the liquid. Due to evaporation, at the end of every year the amount of liquid in the barrel is reduced by 15% of its volume at the start of the year.

    a Calculate the amount of liquid in the barrel at the end of the first year.

    b Show that the amount of liquid in the barrel at the end of ten years is approximately 31.5 litres.

    At the start of each year a new barrel is filled with 160 litres of liquid so that, at the end of 20 years, there are 20 barrels containing liquid.

    c Calculate the total amount of liquid, to the nearest litre, in the barrels at the end of 20 years.

15 At the beginning of the year 2000 a company bought a new machine for £15 000. Each year the value of the machine decreases by 20% of its value at the start of the year.

    a Show that at the start of the year 2002, the value of the machine was £9600.

    b When the value of the machine falls below £500, the company will replace it. Find the year in which the machine will be replaced.

    c To plan for a replacement machine, the company pays £1000 at the start of each year into a savings account. The account pays interest of 5% per annum. The first payment was made when the machine was first bought and the last payment will be made at the start of the year in which the machine is replaced. Using your answer to part **b**, find how much the savings account will be worth when the machine is replaced. *E*

16 A mortgage is taken out for £80 000. It is to be paid by annual instalments of £5000 with the first payment being made at the end of the first year that the mortgage was taken out. Interest of 4% is then charged on any outstanding debt. Find the total time taken to pay off the mortgage.

> **Hint for question 16:** Find an expression for the debt remaining after n years and solve using the fact that if it is paid off, the debt = 0.

Summary of key points

1 In a geometric series you get from one term to the next by multiplying by a constant called the common ratio.

2 The formula for the nth term $= ar^{n-1}$ where $a =$ first term and $r =$ common ratio.

3 The formula for the sum to n terms is
$$S_n = \frac{a(1-r^n)}{1-r} \text{ or } S_n = \frac{a(r^n-1)}{r-1}$$

4 The sum to infinity exists if $|r| < 1$ and is $S_\infty = \dfrac{a}{1-r}$

Graphs of trigonometric functions

In this chapter you will learn about the sine, cosine and tangent functions and their graphs.

You need to be able to use the three basic trigonometric functions for any angle.

You can use the trigonometric ratios

$$\sin\theta = \frac{\text{opposite}}{\text{hypotenuse}} \qquad \cos\theta = \frac{\text{adjacent}}{\text{hypotenuse}} \qquad \tan\theta = \frac{\text{opposite}}{\text{adjacent}}$$

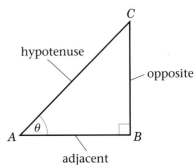

to find the missing sides and angles in a right-angled triangle.

To extend the work on sine, cosine and tangents to cover angles of any size, both positive and negative, we need to modify these definitions.

You need to know what is meant by positive and negative angles.

Example 1

The line OP, where O is the origin, makes an angle θ with the positive x-axis. Draw diagrams to show the position of OP where θ equals:

a $+60°$ **b** $+210°$ **c** $-60°$ **d** $-200°$

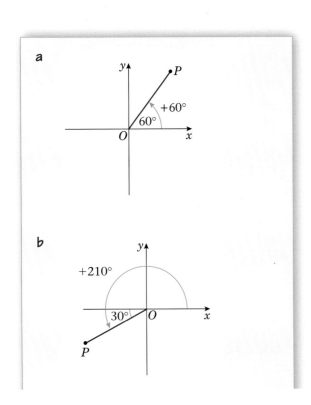

c

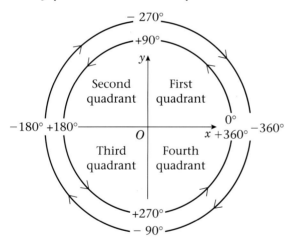

You could also give the angles in radians as

$$\textbf{a}\ +\frac{\pi}{3}\quad \textbf{b}\ +\frac{7\pi}{6}\quad \textbf{c}\ -\frac{\pi}{3}\quad \textbf{d}\ -\frac{10\pi}{9}$$

d

Remember: Anticlockwise angles are positive, clockwise angles are negative, measured from the positive x-axis.

■ **The x–y plane is divided into quadrants:**

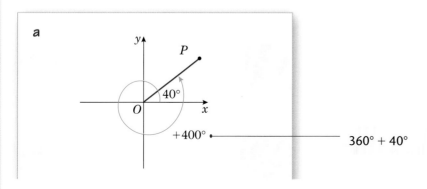

Hint: Angles may lie outside the range 0–360°, but they will always lie in one of the four quadrants.

Example 2

Draw diagrams to show the position of OP where $\theta =$: **a** $+400°$, **b** $+700°$, **c** $-480°$.

a

$360° + 40°$

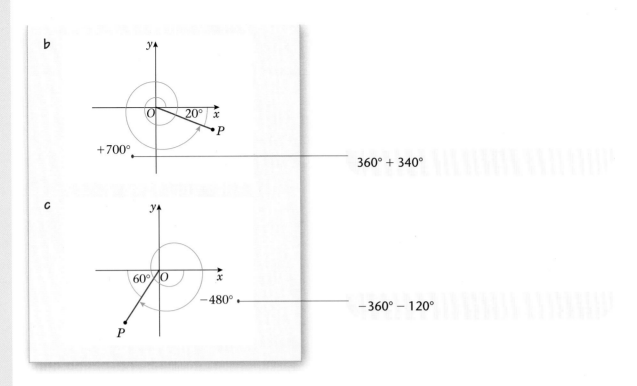

b

$+700°$

$360° + 340°$

c

$-480°$

$-360° - 120°$

Exercise 8A

1 Draw diagrams, as in Examples **1** and **2**, to show the following angles. Mark in the acute angle that OP makes with the x-axis.

a $-80°$	**b** $100°$	**c** $200°$
d $165°$	**e** $-145°$	**f** $225°$
g $280°$	**h** $330°$	**i** $-160°$
j $-280°$	**k** $\dfrac{3\pi}{4}$	**l** $\dfrac{7\pi}{6}$
m $-\dfrac{5\pi}{3}$	**n** $-\dfrac{5\pi}{8}$	**o** $\dfrac{19\pi}{9}$

2 State the quadrant that OP lies in when the angle that OP makes with the positive x-axis is:

a $400°$	**b** $115°$	**c** $-210°$	**d** $255°$
e $-100°$	**f** $\dfrac{7\pi}{8}$	**g** $-\dfrac{11\pi}{6}$	**h** $\dfrac{13\pi}{7}$

■ For all values of θ, the definitions of $\sin\theta$, $\cos\theta$ and $\tan\theta$ are taken to be

$$\sin\theta = \frac{y}{r} \quad \cos\theta = \frac{x}{r} \quad \tan\theta = \frac{y}{x}$$

where x and y are the coordinates of P and r is the length of OP.

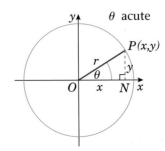
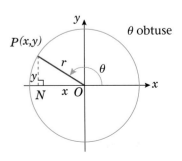

The values of $\sin\theta$ and $\cos\theta$, where θ is a multiple of $90°$, follow from the definitions above.

Example 3

Write down the values of **a** sin 90°, **b** sin 180°, **c** sin 270°, **d** cos 180°, **e** cos (−90)° **f** cos 450°.

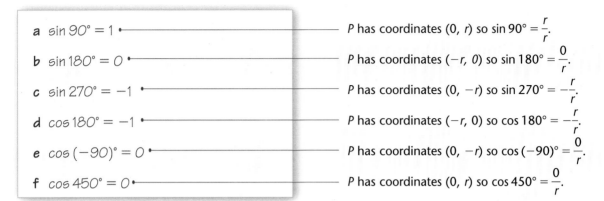

a $\sin 90° = 1$ —— P has coordinates (0, r) so $\sin 90° = \dfrac{r}{r}$.

b $\sin 180° = 0$ —— P has coordinates (−r, 0) so $\sin 180° = \dfrac{0}{r}$.

c $\sin 270° = -1$ —— P has coordinates (0, −r) so $\sin 270° = -\dfrac{r}{r}$.

d $\cos 180° = -1$ —— P has coordinates (−r, 0) so $\cos 180° = -\dfrac{r}{r}$.

e $\cos (-90)° = 0$ —— P has coordinates (0, −r) so $\cos (-90)° = \dfrac{0}{r}$.

f $\cos 450° = 0$ —— P has coordinates (0, r) so $\cos 450° = \dfrac{0}{r}$.

Tan $\theta = \dfrac{y}{x}$ so when $x = 0$ and $y \neq 0$ tan θ is indeterminate.

This is when P is at (0, r) or (0, −r).

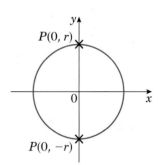

- Tan θ is indeterminate when θ is an odd multiple of 90° $\left(\text{or } \dfrac{\pi}{2} \text{ radians}\right)$.

When $y = 0$, tan $\theta = 0$. This is when P is at (r, 0) or (−r, 0).

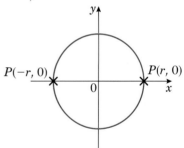

- Tan $\theta = 0$ when θ is 0° or an even multiple of 90° $\left(\text{or } \dfrac{\pi}{2} \text{ radians}\right)$.

Exercise 8B

(*Note*: do not use a calculator.)

1 Write down the values of:

  a sin (−90)°   **b** sin 450°   **c** sin 540°   **d** sin (−450)°

  e cos (−180)°   **f** cos (−270)°   **g** cos 270°   **h** cos 810°

  i tan 360°   **j** tan (−180)°

2 Write down the values of the following, where the angles are in radians:

  a $\sin \dfrac{3\pi}{2}$ **b** $\sin \left(-\dfrac{\pi}{2}\right)$ **c** $\sin 3\pi$ **d** $\sin \dfrac{7\pi}{2}$

  e $\cos 0$ **f** $\cos \pi$ **g** $\cos \dfrac{3\pi}{2}$ **h** $\cos \left(-\dfrac{3\pi}{2}\right)$

  i $\tan \pi$ **j** $\tan (-2\pi)$

8.2 You need to know the signs of the three trigonometric functions in the four quadrants.

In the first quadrant sin, cos and tan are positive.

By considering the sign of x and y, the coordinates of P, you can find the sign of the three trigonometric functions in the other quadrants.

Example 4

Find the signs of $\sin \theta$, $\cos \theta$ and $\tan \theta$ in the second quadrant (θ is obtuse, $90° < \theta < 180°$).

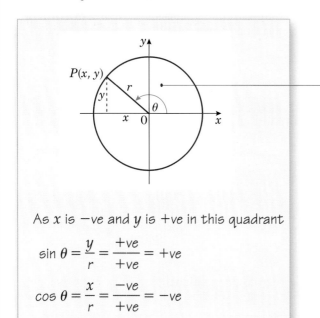

Draw a circle, centre O and radius r, with $P(x, y)$ on the circle in the second quadrant.

As x is $-$ve and y is $+$ve in this quadrant

$$\sin \theta = \frac{y}{r} = \frac{+ve}{+ve} = +ve$$

$$\cos \theta = \frac{x}{r} = \frac{-ve}{+ve} = -ve$$

$$\tan \theta = \frac{y}{x} = \frac{+ve}{-ve} = -ve$$

So only $\sin \theta$ is positive.

■ In the *first* quadrant sin, cos and tan are *all* +ve.
   In the *second* quadrant only *sine* is +ve.
   In the *third* quadrant only *tan* is +ve.
   In the *fourth* quadrant only *cos* is +ve.

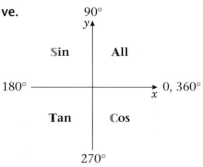

The diagram shows which trigonometric functions are *positive* in each quadrant.

You might find it useful to make up a mnemonic to remember these results.
For example: **A**ll **S**ilver **T**oy **C**ars.

If you make one up, it is a good idea to keep to the order **A**, **S**, **T**, **C**.

Example 5

Show that:

a $\sin (180 - \theta)° = \sin \theta°$

b $\sin (180 + \theta)° = -\sin \theta°$

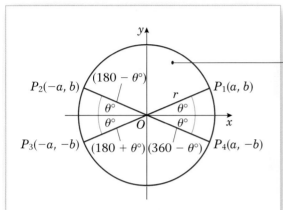

As $\sin \theta = \dfrac{y}{r}$, it follows that:

$$\sin (180 - \theta)° = \frac{b}{r} = \sin \theta°$$

$$\sin (180 + \theta)° = -\frac{b}{r} = -\sin \theta°$$

$$\sin (360 - \theta)° = \frac{-b}{r} = -\sin \theta°$$

Draw a diagram to show the position of the four angles $\theta°$, $(180 - \theta)°$, $(180 + \theta)°$ and $(360 - \theta)°$.

Hint: The four lines, OP_1, OP_2, OP_3 and OP_4, representing the four angles are all inclined at $\theta°$ to the horizontal.

■ **The results for sine, cosine and tangent are:**

$\sin (180 - \theta)° = \sin \theta°$

$\sin (180 + \theta)° = -\sin \theta°$

$\sin (360 - \theta)° = -\sin \theta°$

$\cos (180 - \theta)° = -\cos \theta°$

$\cos (180 + \theta)° = -\cos \theta°$

$\cos (360 - \theta)° = \cos \theta°$

$\tan (180 - \theta)° = -\tan \theta°$

$\tan (180 + \theta)° = \tan \theta°$

$\tan (360 - \theta)° = -\tan \theta°$

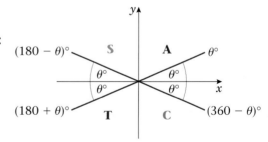

Hint: For angles measured in radians, the same results hold, with 180° being replaced by π, e.g. $\sin (\pi - \theta) = \sin \theta$; $\cos (\pi + \theta) = -\cos \theta$; $\tan (2\pi - \theta) = -\tan \theta$.

Hint: All angles that are equally inclined to either the +ve x-axis or the −ve x-axis have trigonometric ratios which are equal in magnitude, but they take the sign indicated by the quadrant the angle is in.

Example 6

Express in terms of trigonometric ratios of acute angles:

a $\sin(-100)°$ **b** $\cos 330°$ **c** $\tan 500°$

a

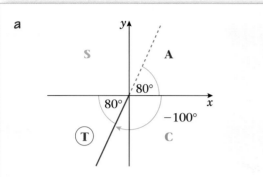

The acute angle made with x-axis is 80°.

In the third quadrant only tan is +ve, so sin is −ve.

So $\sin(-100)° = -\sin 80°$

For each part, draw diagrams showing the position of *OP* for the given angle and insert the acute angle that OP makes with the *x*-axis.

b

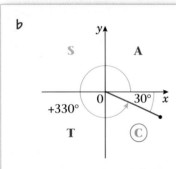

The acute angle made with x-axis is 30°.

In the fourth quadrant only cos is +ve.

So $\cos 330° = +\cos 30°$

c

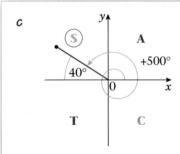

The acute angle made with x-axis is 40°.

In the second quadrant only sin is +ve.

So $\tan 500° = -\tan 40°$

Exercise 8C

(*Note*: Do not use a calculator.)

1 By drawing diagrams, as in Example **6**, express the following in terms of trigonometric ratios of acute angles:

a $\sin 240°$	**b** $\sin (-80)°$	**c** $\sin (-200)°$	**d** $\sin 300°$	**e** $\sin 460°$
f $\cos 110°$	**g** $\cos 260°$	**h** $\cos (-50)°$	**i** $\cos (-200)°$	**j** $\cos 545°$
k $\tan 100°$	**l** $\tan 325°$	**m** $\tan (-30)°$	**n** $\tan (-175)°$	**o** $\tan 600°$
p $\sin \dfrac{7\pi}{6}$	**q** $\cos \dfrac{4\pi}{3}$	**r** $\cos \left(-\dfrac{3\pi}{4}\right)$	**s** $\tan \dfrac{7\pi}{5}$	**t** $\tan \left(-\dfrac{\pi}{3}\right)$
u $\sin \dfrac{15\pi}{16}$	**v** $\cos \dfrac{8\pi}{5}$	**w** $\sin \left(-\dfrac{6\pi}{7}\right)$	**x** $\tan \dfrac{15\pi}{8}$	

2 Given that θ is an acute angle measured in degrees, express in terms of $\sin \theta$:

a $\sin (-\theta)$	**b** $\sin (180° + \theta)$	**c** $\sin (360° - \theta)$
d $\sin -(180° + \theta)$	**e** $\sin (-180° + \theta)$	**f** $\sin (-360° + \theta)$
g $\sin (540° + \theta)$	**h** $\sin (720° - \theta)$	**i** $\sin (\theta + 720°)$

3 Given that θ is an acute angle measured in degrees, express in terms of $\cos \theta$ or $\tan \theta$:

a $\cos (180° - \theta)$	**b** $\cos (180° + \theta)$	**c** $\cos (-\theta)$
d $\cos -(180° - \theta)$	**e** $\cos (\theta - 360°)$	**f** $\cos (\theta - 540°)$
g $\tan (-\theta)$	**h** $\tan (180° - \theta)$	**i** $\tan (180° + \theta)$
j $\tan (-180° + \theta)$	**k** $\tan (540° - \theta)$	**l** $\tan (\theta - 360°)$

> The results obtained in questions **2** and **3** are true for all values of θ.

4 A function f is an even function if $f(-\theta) = f(\theta)$.

A function f is an odd function if $f(-\theta) = -f(\theta)$.

Using your results from questions **2a**, **3c** and **3g**, state whether $\sin \theta$, $\cos \theta$ and $\tan \theta$ are odd or even functions.

8.3 **You need to be able to find the exact values of some trigonometrical ratios.**

You can find the trigonometrical ratios of angles 30°, 45° and 60° exactly.
Consider an equilateral triangle ABC of side 2 units.
If you drop a perpendicular from A to meet BC at D,
then $BD = DC = 1$ unit, $\angle BAD = 30°$ and $\angle ABD = 60°$.
Using Pythagoras' theorem in $\triangle ABD$

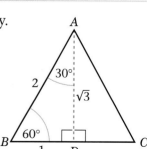

$$AD^2 = 2^2 - 1^2 = 3$$
So $AD = \sqrt{3}$ units

Using $\triangle ABD$, $\sin 30° = \dfrac{1}{2}$, $\cos 30° = \dfrac{\sqrt{3}}{2}$, $\tan 30° = \dfrac{1}{\sqrt{3}} = \dfrac{\sqrt{3}}{3}$,

and $\sin 60° = \dfrac{\sqrt{3}}{2}$, $\cos 60° = \dfrac{1}{2}$, $\tan 60° = \sqrt{3}$.

If you now consider an isosceles right-angled triangle PQR, in which $PQ = QR = 1$ unit, then the ratios for 45° can be found.

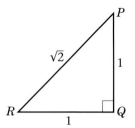

Using Pythagoras' theorem

$$PR^2 = 1^2 + 1^2 = 2$$

So $\quad PR = \sqrt{2}$ units

Then $\sin 45° = \cos 45° = \dfrac{1}{\sqrt{2}} = \dfrac{\sqrt{2}}{2}$ and $\tan 45° = 1$

Exercise 8D

1 Express the following as trigonometric ratios of either 30°, 45° or 60°, and hence find their exact values.

    a $\sin 135°$ **b** $\sin(-60°)$ **c** $\sin 330°$ **d** $\sin 420°$ **e** $\sin(-300°)$

    f $\cos 120°$ **g** $\cos 300°$ **h** $\cos 225°$ **i** $\cos(-210°)$ **j** $\cos 495°$

    k $\tan 135°$ **l** $\tan(-225°)$ **m** $\tan 210°$ **n** $\tan 300°$ **o** $\tan(-120°)$

2 In Section 8.3 you saw that $\sin 30° = \cos 60°$, $\cos 30° = \sin 60°$, and $\tan 60° = \dfrac{1}{\tan 30°}$.

These are particular examples of the general results: $\sin(90° - \theta) = \cos \theta$, and

$\cos(90° - \theta) = \sin \theta$, and $\tan(90° - \theta) = \dfrac{1}{\tan \theta}$, where the angle θ is measured in degrees.

Use a right-angled triangle ABC to verify these results for the case when θ is acute.

8.4 You need to be able to recognise the graphs of $\sin \theta$, $\cos \theta$ and $\tan \theta$.

$y = \sin \theta$

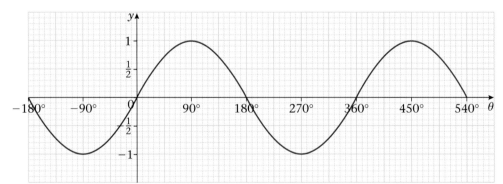

Functions that repeat themselves after a certain interval are called periodic functions, and the interval is called the period of the function. You can see that $\sin \theta$ is periodic with a period of 360°.

> **Hint:** The graph of $\sin \theta$, where θ is in radians, has period 2π.

There are many symmetry properties of $\sin \theta$ (some were seen in Example 5) but you can see from the graph that

$\sin(\theta + 360°) = \sin \theta°$ and $\sin(\theta - 360°) = \sin \theta$

$\sin(90° - \theta) = \sin(90° + \theta)$

> **Hint:** Because it is periodic.

> **Hint:** Symmetry about $\theta = 90°$.

$y = \cos \theta$

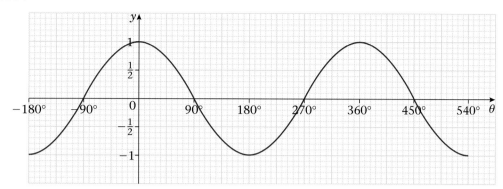

Like sin θ, cos θ is periodic with a period of 360°. In fact, the graph of cos θ is the same as that of sin θ when it has been translated by 90° to the left.

Two further symmetry properties of cos θ are

> Hint: Because it is periodic.

$$\cos(\theta + 360°) = \cos \theta \text{ and } \cos(\theta - 360°) = \cos \theta$$
$$\cos(-\theta) = \cos \theta \text{ (seen on page 117)}$$

> Hint: Symmetry about $\theta = 0°$.

$y = \tan \theta$

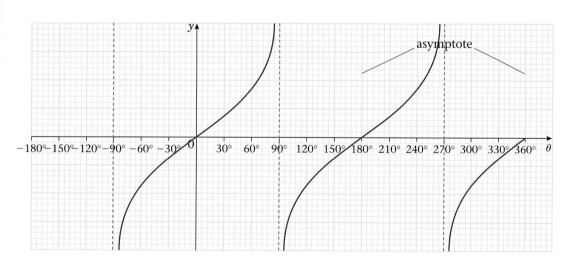

This function behaves very differently from the sine and cosine functions but it is still periodic, it repeats itself in cycles of 180° so its period is 180°.

The period symmetry properties of tan θ are

$$\tan(\theta + 180°) = \tan \theta$$
$$\tan(\theta - 180°) = \tan \theta$$

> Hint: The dotted lines on the graph are called asymptotes, lines to which the curve approaches but never reaches; these occur at $\theta = (2n + 1)90°$ where n is integer.

Example 7

Sketch the graph of $y = \cos \theta°$ in the interval $-360 \leqslant \theta \leqslant 360$.

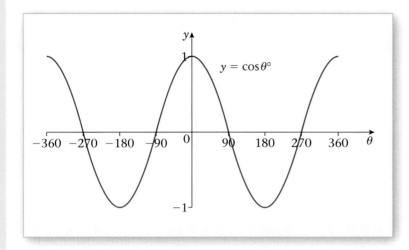

The axes are θ and y.
The curves meets the θ-axis at $\theta = \pm 270°$ and $\theta = \pm 90°$.
Note that the form of the equation given here means that θ is a number.
The curve crosses the y-axis at (0, 1).

Example 8

Sketch the graph of $y = \sin x$ in the interval $-\pi \leqslant x \leqslant \dfrac{3\pi}{2}$.

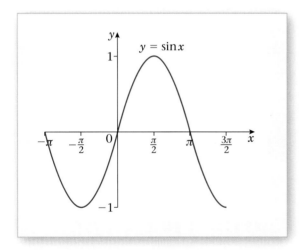

Here the axes are x and y, and the interval tells you that x is in radians.
The curve meets the x-axis at $x = \pm \pi$ and $x = 0$.

Exercise 8E

1 Sketch the graph of $y = \cos \theta$ in the interval $-\pi \leqslant \theta \leqslant \pi$.

2 Sketch the graph of $y = \tan \theta°$ in the interval $-180 \leqslant \theta \leqslant 180$.

3 Sketch the graph of $y = \sin \theta°$ in the interval $-90 \leqslant \theta \leqslant 270$.

8.5 You need to be able to perform simple transformations on the graphs of sin θ, cos θ and tan θ.

You can sketch the graphs of $a \sin \theta$, $a \cos \theta$, and $a \tan \theta$, where a is a constant.

Example 9

Sketch on separate axes the graphs of:

a $y = 3\sin x$, $0 \leqslant x \leqslant 360°$

b $y = -\tan \theta$, $-\pi \leqslant \theta \leqslant \pi$

a

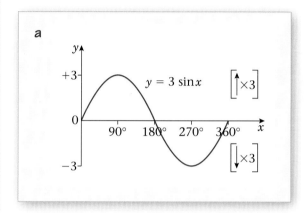

The effect of the multiplication factor 3, is to stretch vertically the graph of sin x by a scale factor of 3; there is no effect in the x-direction.

In this case the labelling of the intercepts on the x-axis are in degrees (°).

b

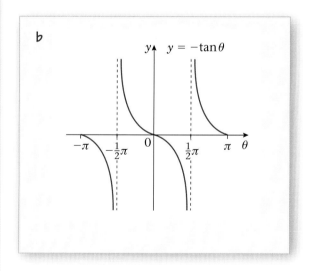

The effect of the multiplication factor -1, is to reflect the graph of tan θ in the θ-axis. Labelling on the θ-axis is in radians.

You can sketch the graphs of sin $\theta + a$, cos $\theta + a$ and tan $\theta + a$.

Example 10

Sketch on separate axes the graphs of:

a $y = -1 + \sin x$, $0 \leqslant x \leqslant 2\pi$ **b** $y = \frac{1}{2} + \cos \theta$, $0 \leqslant \theta \leqslant 360°$

a

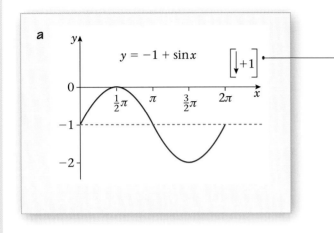

The graph of $y = \sin x$ is translated by 1 unit in the negative y-direction.

b

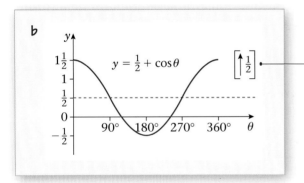

The graph of $y = \cos \theta$ is translated by $\frac{1}{2}$ unit in the positive y-direction. To find the intercepts on the θ-axis requires solving the equation $\frac{1}{2} + \cos \theta = 0$, but if you look back to the graph on page 119 you can see that these are at 120° and 240°.

You can sketch the graphs of sin $(\theta + \alpha)$, cos $(\theta + \alpha)$ and tan $(\theta + \alpha)$.

Example 11

Sketch on separate axes the graphs of:

a $y = \tan\left(\theta + \dfrac{\pi}{4}\right)$, $0 \leqslant \theta \leqslant 2\pi$ **b** $y = \cos(\theta - 90°)$, $-360° \leqslant \theta \leqslant 360°$

a

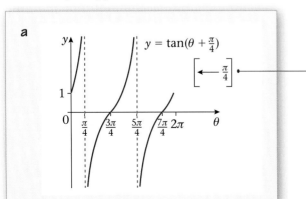

The graph of $y = \tan \theta$ is translated by $\dfrac{\pi}{4}$ to the left. The asymptotes are now at $\theta = \dfrac{\pi}{4}$ and $\theta = \dfrac{5\pi}{4}$. The curve meets the y-axis where $\theta = 0$, so $y = 1$.

b

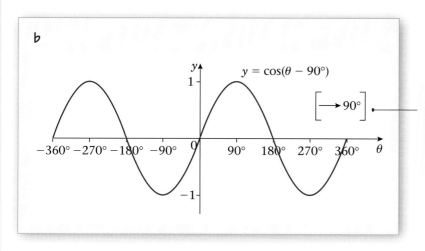

$y = \cos(\theta - 90°)$

The graph of $y = \cos\theta$ is translated by 90° to the right. Note that this is exactly the same curve as $y = \sin\theta$, so another property is that $\cos(\theta - 90°) = \sin\theta$.

You can sketch the graphs of sin $n\theta$, cos $n\theta$ and tan $n\theta$.

Remember that the curve with equation $y = \text{f}(ax)$ is a horizontal stretch with scale factor $\dfrac{1}{a}$ of the curve $y = \text{f}(x)$. In the special case where $a = -1$, this is equivalent to a reflection in the y-axis.

Example 12

Sketch on separate axes the graphs of:

a $y = \sin 2x$, $0 \le x \le 360°$

b $y = \cos\dfrac{\theta}{3}$, $-3\pi \le \theta \le 3\pi$

c $y = \tan(-x)$, $-360° \le x \le 360°$

a

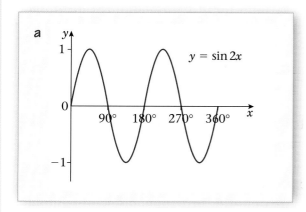

$y = \sin 2x$

The graph of $y = \sin x$ is stretched horizontally with scale factor $\frac{1}{2}$.

The period is now 180° and two complete 'waves' are seen in the interval $0 \le x \le 360°$.

b

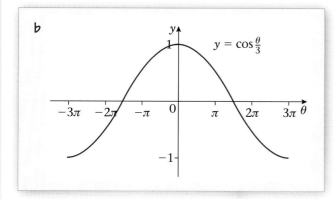

$y = \cos\dfrac{\theta}{3}$

The graph of $y = \cos\theta$ is stretched horizontally with scale factor 3.

The period of $\cos\dfrac{\theta}{3}$ is 6π and only one complete wave is seen in $-3\pi \le \theta \le 3\pi$.

The curve crosses the θ-axis at $\theta = \pm\frac{3}{2}\pi$.

c

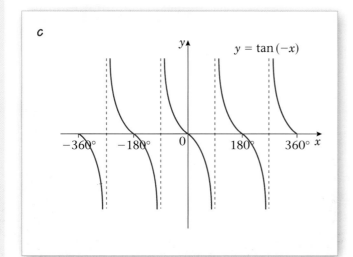

$y = \tan(-x)$

The graph of $y = \tan x$ is reflected in the y-axis.

Exercise 8F

1   Write down **i** the maximum value, and **ii** the minimum value, of the following expressions, and in each case give the smallest positive (or zero) value of x for which it occurs.

   a $\cos x°$ **b** $4 \sin x°$ **c** $\cos(-x)°$

   d $3 + \sin x°$ **e** $-\sin x°$ **f** $\sin 3x°$

2   Sketch, on the same set of axes, in the interval $0 \leqslant \theta \leqslant 360°$, the graphs of $\cos \theta$ and $\cos 3\theta$.

3   Sketch, on separate axes, the graphs of the following, in the interval $0 \leqslant \theta \leqslant 360°$. Give the coordinates of points of intersection with the axes, and of maximum and minimum points where appropriate.

   a $y = -\cos \theta$ **b** $y = \frac{1}{3} \sin \theta$

   c $y = \sin \frac{1}{3}\theta$ **d** $y = \tan(\theta - 45°)$

4   Sketch, on separate axes, the graphs of the following, in the interval $-180 \leqslant \theta \leqslant 180$. Give the coordinates of points of intersection with the axes, and of maximum and minimum points where appropriate.

   a $y = -2 \sin \theta°$ **b** $y = \tan(\theta + 180)°$ **c** $y = \cos 4\theta°$ **d** $y = \sin(-\theta)°$

5   In this question θ is measured in radians. Sketch, on separate axes, the graphs of the following in the interval $-2\pi \leqslant \theta \leqslant 2\pi$. In each case give the periodicity of the function.

   a $y = \sin \frac{1}{2}\theta$ **b** $y = -\frac{1}{2}\cos \theta$ **c** $y = \tan\left(\theta - \dfrac{\pi}{2}\right)$ **d** $y = \tan 2\theta$

6   **a** By considering the graphs of the functions, or otherwise, verify that:

   i $\cos \theta = \cos(-\theta)$

   ii $\sin \theta = -\sin(-\theta)$

   iii $\sin(\theta - 90°) = -\cos \theta$

   b Use the results in **a ii** and **iii** to show that $\sin(90° - \theta) = \cos \theta$.

   c In Example 11 you saw that $\cos(\theta - 90°) = \sin \theta$.
       Use this result with part **a i** to show that $\cos(90° - \theta) = \sin \theta$.

Mixed exercise 8G

1 Write each of the following as a trigonometric ratio of an acute angle:

    a $\cos 237°$ **b** $\sin 312°$ **c** $\tan 190°$ **d** $\sin 2.3^c$ **e** $\cos\left(-\dfrac{\pi}{15}\right)$

2 Without using your calculator, work out the values of:

    a $\cos 270°$ **b** $\sin 225°$ **c** $\cos 180°$ **d** $\tan 240°$ **e** $\tan 135°$

    f $\cos 690°$ **g** $\sin \dfrac{5\pi}{3}$ **h** $\cos\left(-\dfrac{2\pi}{3}\right)$ **i** $\tan 2\pi$ **j** $\sin\left(-\dfrac{7\pi}{6}\right)$

3 Describe geometrically the transformations which map:

    a The graph of $y = \tan x°$ onto the graph of $\tan \frac{1}{2}x°$.

    b The graph of $y = \tan \frac{1}{2}x°$ onto the graph of $3 + \tan \frac{1}{2}x°$.

    c The graph of $y = \cos x°$ onto the graph of $-\cos x°$.

    d The graph of $y = \sin (x - 10)°$ onto the graph of $\sin (x + 10)°$.

4 **a** Sketch on the same set of axes, in the interval $0 \le x \le \pi$, the graphs of $y = \tan (x - \frac{1}{4}\pi)$ and $y = -2\cos x$, showing the coordinates of points of intersection with the axes.

    b Deduce the number of solutions of the equation $\tan (x - \frac{1}{4}\pi) + 2\cos x = 0$, in the interval $0 \le x \le \pi$.

5 The diagram shows part of the graph of $y = f(x)$. It crosses the x-axis at $A(120, 0)$ and $B(p, 0)$. It crosses the y-axis at $C(0, q)$ and has a maximum value at D, as shown.

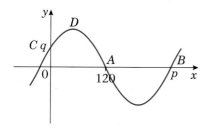

    Given that $f(x) = \sin (x + k)°$, where $k > 0$, write down:

    a the value of p

    b the coordinates of D

    c the smallest value of k

    d the value of q

6 Consider the function $f(x) = \sin px$, $p \in \mathbb{R}$, $0 \le x \le 2\pi$.

    The closest point to the origin that the graph of $f(x)$ crosses the x-axis has x-coordinate $\dfrac{\pi}{5}$.

    a Sketch the graph of $f(x)$.

    b Write down the period of $f(x)$.

    c Find the value of p.

7 The graph below shows $y = \sin\theta$, $0 \leqslant \theta \leqslant 360°$, with one value of θ ($\theta = \alpha°$) marked on the axis.

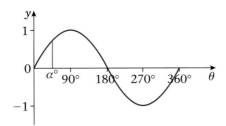

a Copy the graph and mark on the θ-axis the positions of $(180 - \alpha)°$, $(180 + \alpha)°$, and $(360 - \alpha)°$.

b Establish the result $\sin\alpha° = \sin(180 - \alpha)° = -\sin(180 + \alpha)° = -\sin(360 - \alpha)°$.

8 **a** Sketch on separate axes the graphs of $y = \cos\theta$ $(0 \leqslant \theta \leqslant 360°)$ and $y = \tan\theta$ $(0 \leqslant \theta \leqslant 360°)$, and on each θ-axis mark the point $(\alpha°, 0)$ as in question **7**.

b Verify that:

i $\cos\alpha° = -\cos(180 - \alpha)° = -\cos(180 + \alpha)° = \cos(360 - \alpha)°$.

ii $\tan\alpha° = -\tan(180 - \alpha)° = -\tan(180 + \alpha)° = -\tan(360 - \alpha)°$.

Summary of key points

1 The *x*–*y* plane is divided into quadrants:

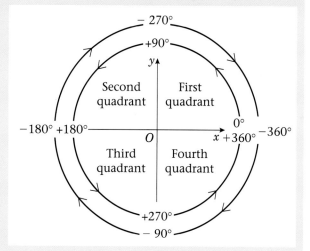

2 For all values of θ, the definitions of $\sin \theta$, $\cos \theta$ and $\tan \theta$ are taken to be

$$\sin \theta = \frac{y}{r} \quad \cos \theta = \frac{x}{r} \quad \tan \theta = \frac{y}{x}$$

where x and y are the coordinates of P and r is the radius of the circle.

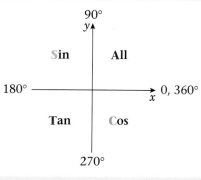

3 In the first quadrant, where θ is acute,
All trigonometric functions are positive.
In the second quadrant, where θ is obtuse, only
Sine is positive.
In the third quadrant, where θ is reflex,
$180° < \theta < 270°$, only **T**angent is positive.
In the fourth quadrant, where θ is reflex,
$270° < \theta < 360°$, only **C**osine is positive.

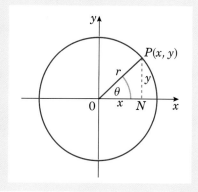

4 The trigonometric ratios of angles equally
inclined to the horizontal are related:
$\sin (180 - \theta)° = \sin \theta°$
$\sin (180 + \theta)° = -\sin \theta°$
$\sin (360 - \theta)° = -\sin \theta°$
$\cos (180 - \theta)° = -\cos \theta°$
$\cos (180 + \theta)° = -\cos \theta°$
$\cos (360 - \theta)° = \cos \theta°$
$\tan (180 - \theta)° = -\tan \theta°$
$\tan (180 + \theta)° = \tan \theta°$
$\tan (360 - \theta)° = -\tan \theta°$

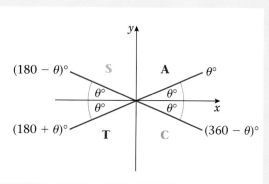

5 The trigonometric ratios of 30°, 45° and 60° have exact forms, given below:

$$\sin 30° = \frac{1}{2} \qquad \cos 30° = \frac{\sqrt{3}}{2} \qquad \tan 30° = \frac{1}{\sqrt{3}} = \frac{\sqrt{3}}{3}$$

$$\sin 45° = \frac{1}{\sqrt{2}} = \frac{\sqrt{2}}{2} \qquad \cos 45° = \frac{1}{\sqrt{2}} = \frac{\sqrt{2}}{2} \qquad \tan 45° = 1$$

$$\sin 60° = \frac{\sqrt{3}}{2} \qquad \cos 60° = \frac{1}{2} \qquad \tan 60° = \sqrt{3}$$

6 The sine and cosine functions have a period of 360°, (or 2π radians). Periodic properties are

$$\sin(\theta \pm 360°) = \sin\theta \text{ and } \cos(\theta \pm 360°) = \cos\theta$$

respectively.

7 The tangent function has a period of 180°, (or π radians).

Periodic properties is $\tan(\theta \pm 180°) = \tan\theta$.

8 Other useful properties are

$$\sin(-\theta) = -\sin\theta; \cos(-\theta) = \cos\theta; \tan(-\theta) = -\tan\theta;$$
$$\sin(90° - \theta) = \cos\theta; \cos(90° - \theta) = \sin\theta$$

9 | Differentiation

In this chapter you will learn to use differentiation so that you can
- distinguish between increasing and decreasing functions
- find the coordinates of maximum and minimum points on a curve
- apply these techniques to real-life problems

9.1 You need to know the difference between increasing and decreasing functions.

A function f which increases as x increases in the interval from $x = a$ to $x = b$ is an increasing function in the interval (a, b).

- **For an increasing function in the interval (a, b), if x_1 and x_2 are two values of x in the interval $a \leqslant x \leqslant b$ and if $x_1 < x_2$ then $f(x_1) < f(x_2)$.**
  It follows that $f'(x) > 0$ in the interval $a \leqslant x \leqslant b$.

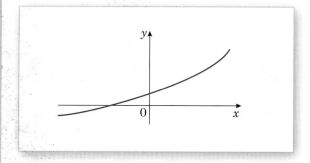

This is a graph of an increasing function. The gradient is positive at all points on the graph.

A function f which decreases as x increases in the interval from $x = a$ to $x = b$ is a decreasing function in the interval (a, b).

- **For a decreasing function in the interval (a, b), if x_1 and x_2 are two values of x in the interval $a \leqslant x \leqslant b$ and if $x_1 < x_2$ then $f(x_1) > f(x_2)$.**
  It follows that $f'(x) < 0$ in the interval $a \leqslant x \leqslant b$.

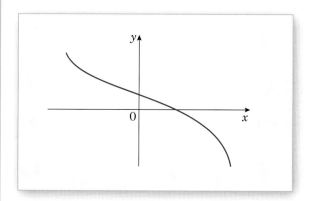

This is a graph of a decreasing function. The gradient is negative at all points on the graph.

Some functions increase as x increases in one interval and decrease as x increases in another interval.

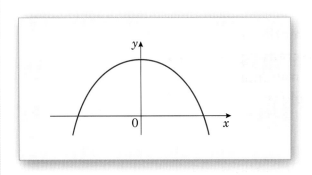

This is the graph of a function which is increasing for $x < 0$, and is decreasing for $x > 0$.
At $x = 0$ the gradient is zero and the function is said to be stationary.

Example 1

Show that the function $f(x) = x^3 + 24x + 3$ $(x \in \Re)$ is an increasing function.

$f(x) = x^3 + 24x + 3$

$f'(x) = 3x^2 + 24$ ←——————— First differentiate to obtain the gradient function.

As $x^2 \geqslant 0$ for all real x

$3x^2 + 24 > 0$

So $f(x)$ is an increasing function. ←——————— As $3x^2 > 0$, for real x, and $24 > 0$ then $f'(x) > 0$ for all values of x. So the curve always has a positive gradient.

Example 2

Find the values of x for which the function $f(x) = x^3 + 3x^2 - 9x$ is a decreasing function.

$f(x) = x^3 + 3x^2 - 9x$

$f'(x) = 3x^2 + 6x - 9$ ——————— Find $f'(x)$ and put this expression < 0.

If $\quad f'(x) < 0 \Rightarrow 3x^2 + 6x - 9 < 0$

So $\quad 3(x^2 + 2x - 3) < 0$ ——————— Solve the inequality by factorisation, and by considering the three regions $x < -3$, $-3 < x < 1$ and $x > 1$, looking for sign changes.

$3(x + 3)(x - 1) < 0$

So $\quad -3 < x < 1$ ←——————— State the answer.

Exercise 9A

1 Find the values of x for which $f(x)$ is an increasing function, given that $f(x)$ equals:

    a $3x^2 + 8x + 2$ **b** $4x - 3x^2$ **c** $5 - 8x - 2x^2$

    d $2x^3 - 15x^2 + 36x$ **e** $3 + 3x - 3x^2 + x^3$ **f** $5x^3 + 12x$

    g $x^4 + 2x^2$ **h** $x^4 - 8x^3$

2 Find the values of x for which f(x) is a decreasing function, given that f(x) equals:

a $x^2 - 9x$ **b** $5x - x^2$ **c** $4 - 2x - x^2$

d $2x^3 - 3x^2 - 12x$ **e** $1 - 27x + x^3$ **f** $x + \dfrac{25}{x}$

g $x^{\frac{1}{2}} + 9x^{-\frac{1}{2}}$ **h** $x^2(x + 3)$

9.2 **You need to be able to find the coordinates of a stationary point on a curve and work out whether it is a minimum point, a maximum point or a point of inflexion.**

For certain curves, the function f(x) is increasing in some intervals and decreasing in others.

■ **The points where f(x) stops increasing and begins to decrease are called maximum points.**

■ **The points where f(x) stops decreasing and begins to increase are called minimum points.**

These points are collectively called turning points and at these points f′(x) = 0. You can find the coordinates of maximum and minimum points on a curve. This will help you to sketch curves accurately.

Some points of inflexion do not have zero gradient.

■ **A point of inflexion is a point where the gradient is at a maximum or minimum value in the neighbourhood of the point.**

> **Hint:** Points of inflexion where the gradient is not zero are not included in the C2 specification.

■ **Points of zero gradient are called stationary points and stationary points may be maximum points, minimum points or points of inflexion.**

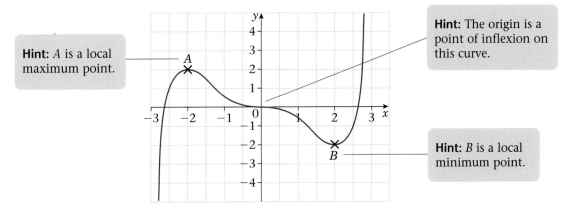

Hint: *A* is a local maximum point.

Hint: The origin is a point of inflexion on this curve.

Hint: *B* is a local minimum point.

■ **To find the coordinates of a stationary point:**

① **Find $\dfrac{dy}{dx}$, i.e. f′(x), and solve the equation f′(x) = 0 to find the value, or values, of x.**

② **Substitute the value(s) of x which you have found into the equation y = f(x) to find the corresponding value(s) of y.**

③ **This gives the coordinates of any stationary points.**

Example 3

Find the coordinates of the turning point on the curve with equation $y = x^4 - 32x$. Establish whether it is a maximum or a minimum point or a point of inflexion by considering points either side of the turning points.

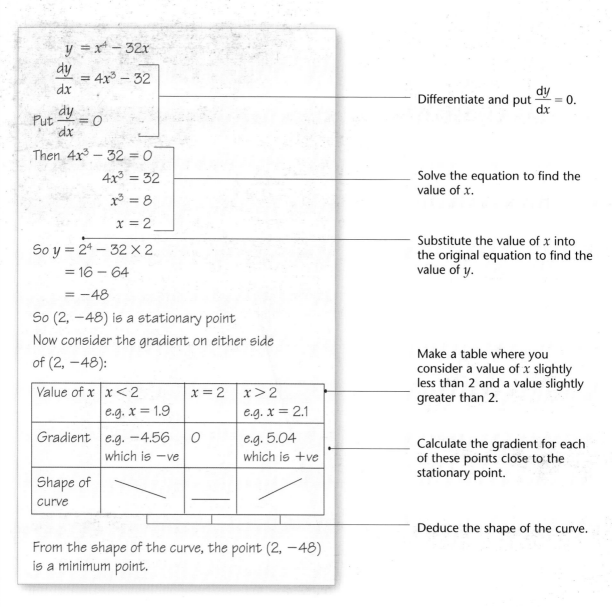

$y = x^4 - 32x$

$\dfrac{dy}{dx} = 4x^3 - 32$ ⟶ Differentiate and put $\dfrac{dy}{dx} = 0$.

Put $\dfrac{dy}{dx} = 0$

Then $4x^3 - 32 = 0$

$4x^3 = 32$ ⟶ Solve the equation to find the value of x.

$x^3 = 8$

$x = 2$

So $y = 2^4 - 32 \times 2$ ⟶ Substitute the value of x into the original equation to find the value of y.

$= 16 - 64$

$= -48$

So $(2, -48)$ is a stationary point

Now consider the gradient on either side of $(2, -48)$:

Value of x	$x < 2$ e.g. $x = 1.9$	$x = 2$	$x > 2$ e.g. $x = 2.1$
Gradient	e.g. -4.56 which is $-$ve	0	e.g. 5.04 which is $+$ve
Shape of curve	\	—	/

Make a table where you consider a value of x slightly less than 2 and a value slightly greater than 2.

Calculate the gradient for each of these points close to the stationary point.

Deduce the shape of the curve.

From the shape of the curve, the point $(2, -48)$ is a minimum point.

■ You can also find out whether stationary points are maximum points, minimum points or points of inflexion by finding the value of $\dfrac{d^2y}{dx^2}$ and, where necessary, $\dfrac{d^3y}{dx^3}$ at the stationary point. This is because $\dfrac{d^2y}{dx^2}$ measures the change in gradient.

• If $\dfrac{dy}{dx} = 0$ and $\dfrac{d^2y}{dx^2} > 0$, the point is a minimum point.

Hint: $\dfrac{d^2y}{dx^2}$ is the second derivative of y with respect to x. You find $\dfrac{d^2y}{dx^2}$ by differentiating $\dfrac{dy}{dx}$ with respect to x.

- If $\dfrac{dy}{dx} = 0$ and $\dfrac{d^2y}{dx^2} < 0$, the point is a maximum point.

- If $\dfrac{dy}{dx} = 0$ and $\dfrac{d^2y}{dx^2} = 0$, the point is either a maximum or a minimum point or a point of inflexion.

Hint: In this case, you need to use the tabular method and consider the gradient on either side of the stationary point.

- If $\dfrac{dy}{dx} = 0$ and $\dfrac{d^2y}{dx^2} = 0$, but $\dfrac{d^3y}{dx^3} \neq 0$, then the point is a point of inflexion.

Hint: $\dfrac{d^3y}{dx^3}$ is the third derivative of y with respect to x. You find $\dfrac{d^3y}{dx^3}$ by differentiating $\dfrac{d^2y}{dx^2}$ with respect to x.

You may also see this notation: f″(x) is the second derivative of the function f with respect to x. f‴(x) is the third derivative and so on.

Example 4

Find the stationary points on the curve with equation $y = 2x^3 - 15x^2 + 24x + 6$ and determine, by finding the second derivative, whether the stationary points are maximum, minimum or points of inflexion.

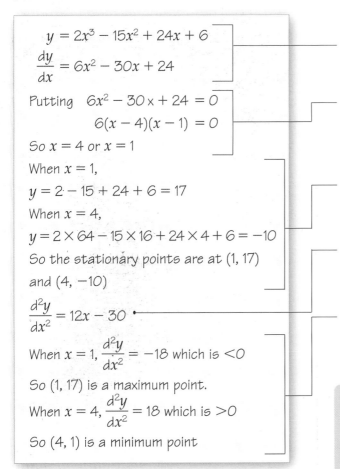

$y = 2x^3 - 15x^2 + 24x + 6$

$\dfrac{dy}{dx} = 6x^2 - 30x + 24$

Differentiate and put the derivative equal to zero.

Putting $6x^2 - 30x + 24 = 0$

$6(x - 4)(x - 1) = 0$

Solve the equation to obtain the values of x for the stationary points.

So $x = 4$ or $x = 1$

When $x = 1$,

$y = 2 - 15 + 24 + 6 = 17$

When $x = 4$,

$y = 2 \times 64 - 15 \times 16 + 24 \times 4 + 6 = -10$

Substitute $x = 4$ and $x = 1$ into the original equation of the curve to obtain the values of y which correspond to these values.

So the stationary points are at (1, 17) and (4, −10)

$\dfrac{d^2y}{dx^2} = 12x - 30$

Differentiate again to obtain the second derivative.

When $x = 1$, $\dfrac{d^2y}{dx^2} = -18$ which is <0

So (1, 17) is a maximum point.

When $x = 4$, $\dfrac{d^2y}{dx^2} = 18$ which is >0

So (4, 1) is a minimum point

Substitute $x = 1$ and $x = 4$ into the second derivative expression. If the second derivative is negative then the point is a maximum point, whereas if it is positive then the point is a minimum point.

Hint: You may be told whether a stationary value is a maximum or a minimum, in which case it will not be necessary for you to check by using the second derivative, or by considering the gradient on each side of the stationary value.

Example 5

Find the greatest value of $6x - x^2$. State the range of the function $f(x) = 6x - x^2$

Let $y = 6x - x^2$

Then $\dfrac{dy}{dx} = 6 - 2x$

Put $\dfrac{dy}{dx} = 0$, then $x = 3$

So $y = 18 - 3^2 = 9$

The greatest value of this quadratic function is 9 and the range is given by $f(x) \le 9$

This question may be done by completing the square, but calculus is a good alternative.

There was only one turning point on this parabola and the question said that there was a greatest value, so you did not need to make a check.

Put the value of x into the original equation. The greatest value is the value of y at the stationary point.

The range of the function is the set of values which y can take.

Exercise 9B

1 Find the least value of each of the following functions:

   a $f(x) = x^2 - 12x + 8$ **b** $f(x) = x^2 - 8x - 1$ **c** $f(x) = 5x^2 + 2x$

2 Find the greatest value of each of the following functions:

   a $f(x) = 10 - 5x^2$ **b** $f(x) = 3 + 2x - x^2$ **c** $f(x) = (6 + x)(1 - x)$

3 Find the coordinates of the points where the gradient is zero on the curves with the given equations. Establish whether these points are maximum points, minimum points or points of inflexion, by considering the second derivative in each case.

   a $y = 4x^2 + 6x$ **b** $y = 9 + x - x^2$ **c** $y = x^3 - x^2 - x + 1$

   d $y = x(x^2 - 4x - 3)$ **e** $y = x + \dfrac{1}{x}$ **f** $y = x^2 + \dfrac{54}{x}$

   g $y = x - 3\sqrt{x}$ **h** $y = x^{\frac{1}{2}}(x - 6)$ **i** $y = x^4 - 12x^2$

4 Sketch the curves with equations given in question **3** parts **a**, **b**, **c** and **d**, labelling any stationary values.

5 By considering the gradient on either side of the stationary point on the curve $y = x^3 - 3x^2 + 3x$, show that this point is a point of inflexion. Sketch the curve $y = x^3 - 3x^2 + 3x$.

6 Find the maximum value and hence the range of values for the function $f(x) = 27 - 2x^4$.

9.3 You need to be able to apply what you have learned about turning points to solve problems.

Example 6

The diagram shows a minor sector *OMN* of a circle with centre *O* and radius *r* cm. The perimeter of the sector is 100 cm and the area of the sector is *A* cm^2.

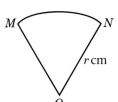

a Show that $A = 50r - r^2$.

Given that *r* varies, find:

b The value of *r* for which *A* is a maximum and show that *A* is a maximum.

c The value of $\angle MON$ for this maximum area.

d The maximum area of the sector *OMN*.

---

a Let the perimeter of the sector be *P*, so

$$P = 2r + r\theta$$

Rearrange and substitute $P = 100$ to give

$$\theta = \frac{100 - 2r}{r} \quad \text{(1)}$$

The area of the sector, $A = \frac{1}{2}r^2\theta \quad \text{(2)}$

Substitute (1) in (2)

$$A = \frac{1}{2}r^2\left(\frac{100 - 2r}{r}\right)$$

So $A = 50r - r^2$

> This is the sum of two radii (2*r*) and an arc length *MN* (*r*θ).

> The area formula is in terms of two variables *r* and θ, so you need to substitute for θ so that the formula is in terms of one variable *r*.

---

b $\dfrac{dA}{dr} = 50 - 2r$

When $\dfrac{dA}{dr} = 0$, $r = 25$

Also $\dfrac{d^2A}{dr^2} = -2$, which is negative

So the area is a maximum when $r = 25$.

> Use the method which you learned to find stationary values: put the first derivative equal to zero, then check the sign of the second derivative.

---

c Substitute $r = 25$ into (1)

$$\theta = \frac{100 - 50}{25} = 2$$

So angle $MON = 2$ radians

> Answer the final two parts of the question by using the appropriate equations and give the units in your answer.

---

d The maximum value of the area is

$$50 \times 25 - 25^2 = 625 \text{ cm}^2$$

> Use $A = 50r - r^2$.

Example 7

A large tank in the shape of a cuboid is to be made from $54\,\text{m}^2$ of sheet metal. The tank has a horizontal base and no top. The height of the tank is x metres. Two of the opposite vertical faces are squares.

a Show that the volume, $V\,\text{m}^3$, of the tank is given by $V = 18x - \frac{2}{3}x^3$.

b Given that x can vary, use differentiation to find the maximum or minimum value of V.

c Justify that the value of V you have found is a maximum.

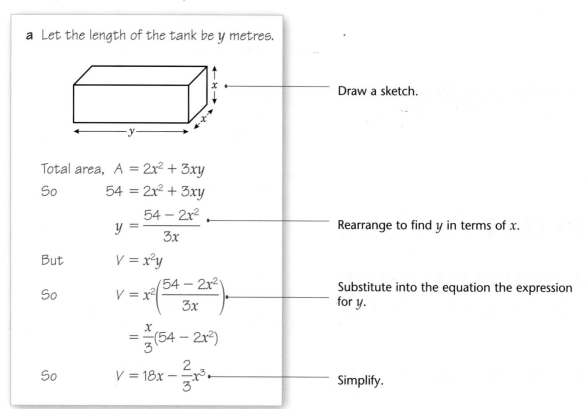

a Let the length of the tank be y metres.

Draw a sketch.

Total area, $A = 2x^2 + 3xy$

So $\qquad 54 = 2x^2 + 3xy$

$$y = \frac{54 - 2x^2}{3x}$$

Rearrange to find y in terms of x.

But $\qquad V = x^2 y$

So $\qquad V = x^2\left(\dfrac{54 - 2x^2}{3x}\right)$

Substitute into the equation the expression for y.

$$= \frac{x}{3}(54 - 2x^2)$$

So $\qquad V = 18x - \dfrac{2}{3}x^3$

Simplify.

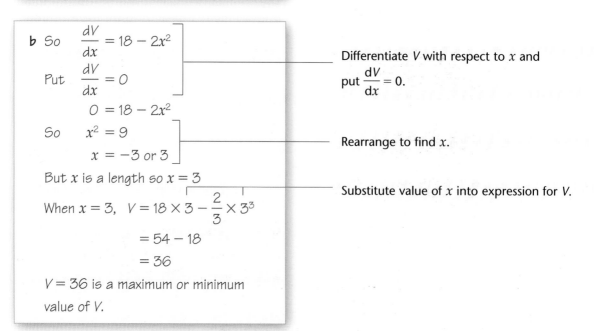

b So $\qquad \dfrac{dV}{dx} = 18 - 2x^2$

Put $\qquad \dfrac{dV}{dx} = 0$

Differentiate V with respect to x and put $\dfrac{dV}{dx} = 0$.

$$0 = 18 - 2x^2$$

So $\qquad x^2 = 9$

$$x = -3 \text{ or } 3$$

Rearrange to find x.

But x is a length so $x = 3$

When $x = 3$, $\quad V = 18 \times 3 - \dfrac{2}{3} \times 3^3$

Substitute value of x into expression for V.

$$= 54 - 18$$

$$= 36$$

$V = 36$ is a maximum or minimum

value of V.

c $\dfrac{d^2V}{dx^2} = -4x$ •———————————— Find the second derivative of V.

When $x = 3$, $\dfrac{d^2V}{dx^2} = -4 \times 3 = -12$

This is negative, so $V = 36$ is the
maximum value of V.

Exercise 9C

1 A rectangular garden is fenced on three sides, and the house forms the fourth side of the rectangle.

Given that the total length of the fence is 80 m show that the area, A, of the garden is given by the formula $A = y(80 - 2y)$, where y is the distance from the house to the end of the garden.

Given that the area is a maximum for this length of fence, find the dimensions of the enclosed garden, and the area which is enclosed.

2 A closed cylinder has total surface area equal to 600π. Show that the volume, $V\,\text{cm}^3$, of this cylinder is given by the formula $V = 300\pi r - \pi r^3$, where $r\,\text{cm}$ is the radius of the cylinder.

Find the maximum volume of such a cylinder.

3 A sector of a circle has area $100\,\text{cm}^2$. Show that the perimeter of this sector is given by the formula $P = 2r + \dfrac{200}{r}$, $r > \sqrt{\dfrac{100}{\pi}}$.

Find the minimum value for the perimeter of such a sector.

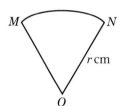

4 A shape consists of a rectangular base with a semicircular top, as shown. Given that the perimeter of the shape is 40 cm, show that its area, $A\,\text{cm}^2$, is given by the formula

$$A = 40r - 2r^2 - \dfrac{\pi r^2}{2}$$

where $r\,\text{cm}$ is the radius of the semicircle. Find the maximum value for this area.

5 The shape shown is a wire frame in the form of a large rectangle split by parallel lengths of wire into 12 smaller equal-sized rectangles.

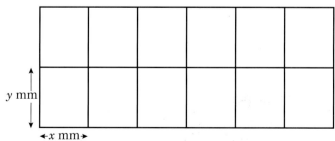

Given that the total length of wire used to complete the whole frame is 1512 mm, show that the area of the whole shape is $A\,\text{mm}^2$, where $A = 1296x - \dfrac{108x^2}{7}$, where x mm is the width of one of the smaller rectangles.

Find the maximum area which can be enclosed in this way.

Mixed exercise 9D

1 Given that: $y = x^{\frac{3}{2}} + \dfrac{48}{x}$ $(x > 0)$

   a Find the value of x and the value of y when $\dfrac{dy}{dx} = 0$.

   b Show that the value of y which you found in **a** is a minimum.

2 A curve has equation $y = x^3 - 5x^2 + 7x - 14$. Determine, by calculation, the coordinates of the stationary points of the curve C.

3 The function f, defined for $x \in R$, $x > 0$, is such that:

$$f'(x) = x^2 - 2 + \frac{1}{x^2}$$

   a Find the value of $f''(x)$ at $x = 4$.

   b Given that $f(3) = 0$, find $f(x)$.

   c Prove that f is an increasing function.

4 A curve has equation $y = x^3 - 6x^2 + 9x$.
Find the coordinates of its maximum turning point.

5 A wire is bent into the plane shape $ABCDEA$ as shown. Shape $ABDE$ is a rectangle and BCD is a semicircle with diameter BD. The area of the region enclosed by the wire is $R\,\text{m}^2$, $AE = x$ metres, $AB = ED = y$ metres. The total length of the wire is $2\,\text{m}$.

   a Find an expression for y in terms of x.

   b Prove that $R = \dfrac{x}{8}(8 - 4x - \pi x)$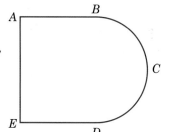

Given that x can vary, using calculus and showing your working,

   c find the maximum value of R. (You do not have to prove that the value you obtain is a maximum.)

6 The fixed point A has coordinates $(8, -6, 5)$ and the variable point P has coordinates $(t, t, 2t)$.

   a Show that $AP^2 = 6t^2 - 24t + 125$.

   b Hence find the value of t for which the distance AP is least.

   c Determine this least distance.

7 A cylindrical biscuit tin has a close-fitting lid which overlaps the tin by 1 cm, as shown. The radii of the tin and the lid are both x cm. The tin and the lid are made from a thin sheet of metal of area $80\pi\,\text{cm}^2$ and there is no wastage. The volume of the tine is $V\,\text{cm}^3$.

   a Show that $V = \pi(40x - x^2 - x^3)$.
Given that x can vary: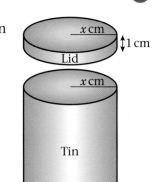

   b Use differentiation to find the positive value of x for which V is stationary.

   c Prove that this value of x gives a maximum value of V.

   d Find this maximum value of V.

   e Determine the percentage of the sheet metal used in the lid when V is a maximum.

8 The diagram shows an open tank for storing water, *ABCDEF*. The sides *ABFE* and *CDEF* are rectangles. The triangular ends *ADE* and *BCF* are isosceles, and $\angle AED = \angle BFC = 90°$. The ends *ADE* and *BCF* are vertical and *EF* is horizontal.

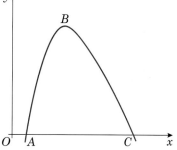

Given that $AD = x$ metres:

a show that the area of triangle *ADE* is $\frac{1}{4}x^2 \, \text{m}^2$.

Given also that the capacity of the container is $4000 \, \text{m}^3$ and that the total area of the two triangular and two rectangular sides of the container is $S \, \text{m}^2$:

b Show that $S = \dfrac{x^2}{2} + \dfrac{16\,000\sqrt{2}}{x}$.

Given that x can vary:

c Use calculus to find the minimum value of S.

d Justify that the value of S you have found is a minimum.

9 The diagram shows part of the curve with equation $y = \text{f}(x)$, where:

$$\text{f}(x) \equiv 200 - \frac{250}{x} - x^2, \ x > 0$$

The curve cuts the x-axis at the points *A* and *C*. The point *B* is the maximum point of the curve.

a Find $\text{f}'(x)$.

b Use your answer to part **a** to calculate the coordinates of *B*.

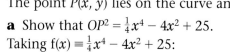

10 The diagram shows the part of the curve with equation $y = 5 - \frac{1}{2}x^2$ for which $y \geqslant 0$. The point $P(x, y)$ lies on the curve and *O* is the origin.

a Show that $OP^2 = \frac{1}{4}x^4 - 4x^2 + 25$.

Taking $\text{f}(x) \equiv \frac{1}{4}x^4 - 4x^2 + 25$:

b Find the values of x for which $\text{f}'(x) = 0$.

c Hence, or otherwise, find the minimum distance from *O* to the curve, showing that your answer is a minimum.

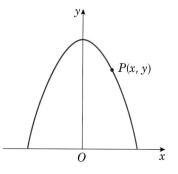

11 The diagram shows part of the curve with equation $y = 3 + 5x + x^2 - x^3$. The curve touches the x-axis at *A* and crosses the x-axis at *C*. The points *A* and *B* are stationary points on the curve.

a Show that *C* has coordinates $(3, 0)$.

b Using calculus and showing all your working, find the coordinates of *A* and *B*.

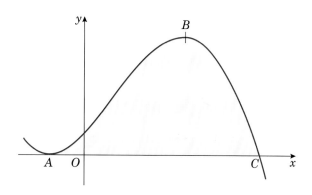

Summary of key points

1 For an increasing function f(x) in the interval (a, b),
   f'(x) > 0 in the interval $a \leqslant x \leqslant b$.

2 For a decreasing function f(x) in the interval (a, b),
   f'(x) < 0 in the interval $a \leqslant x \leqslant b$.

3 The points where f(x) stops increasing and begins to
   decrease are called maximum points.

4 The points where f(x) stops decreasing and begins to
   increase are called minimum points.

5 A point of inflexion is a point where the gradient is at a
   maximum or minimum value in the neighbourhood of
   the point.

6 A stationary point is a point of zero gradient. It may be
   a maximum, a minimum or a point of inflexion.

7 To find the coordinates of a stationary point find
   $\dfrac{dy}{dx}$, i.e. f'(x), and solve the equation f'(x) = 0 to find the
   value, or values, of x and then substitute into y = f(x) to
   find the corresponding values of y.

8 The stationary value of a function is the value of y at
   the stationary point. You can sometimes use this to find
   the range of a function.

9 You may determine the nature of a stationary point by
   using the second derivative.

   If $\dfrac{dy}{dx} = 0$ and $\dfrac{d^2y}{dx^2} > 0$, the point is a minimum point.

   If $\dfrac{dy}{dx} = 0$ and $\dfrac{d^2y}{dx^2} < 0$, the point is a maximum point.

   If $\dfrac{dy}{dx} = 0$ and $\dfrac{d^2y}{dx^2} = 0$, the point is either a maximum
   or a minimum point or a point of inflexion.

   If $\dfrac{dy}{dx} = 0$ and $\dfrac{d^2y}{dx^2} = 0$, but $\dfrac{d^3y}{dx^3} \neq 0$, then the point is a
   point of inflexion.

   Hint: In this case you
   need to use the tabular
   method and consider the
   gradient on either side of
   the stationary point.

10 In problems where you need to find the maximum or
   minimum value of a variable y, first establish a formula
   for y in terms of x, then differentiate and put the
   derived function equal to zero to find x and then y.

10 Trigonometrical identities and simple equations

This chapter introduces trigonometric identities and simple trigonometrical equations.

10.1 You need to be able to use the relationships $\tan \theta \equiv \dfrac{\sin \theta}{\cos \theta}$ and $\sin^2 \theta + \cos^2 \theta \equiv 1$

You saw on page 112 that for all values of θ

$$\sin \theta = \frac{y}{r} \quad \cos \theta = \frac{x}{r} \quad \tan \theta = \frac{y}{x}$$

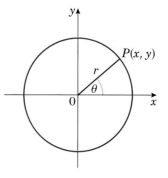

where (x, y) are the coordinates of the point P as it moves round the circumference of a circle with centre O and radius r, and OP makes an angle θ with the +ve x-axis.

Now, $\tan \theta = \dfrac{y}{x} = \dfrac{y}{r} \times \dfrac{r}{x} = \sin \theta \times \dfrac{1}{\cos \theta} = \dfrac{\sin \theta}{\cos \theta}$, so $\tan \theta = \dfrac{\sin \theta}{\cos \theta}$

■ For all values of θ (except where $\cos \theta = 0$, i.e. for odd multiples of 90°, where $\tan \theta$ is not defined):

$$\tan \theta \equiv \frac{\sin \theta}{\cos \theta}$$

You know from Chapter 4 that the equation of the circle with centre the origin and radius r is

$$x^2 + y^2 = r^2$$

Using the equations at the top of the page you can express the coordinates of P in terms of r and θ:

$$x = r \cos \theta \text{ and } y = r \sin \theta$$

Since P lies on this circle:

$$(r \cos \theta)^2 + (r \sin \theta)^2 = r^2$$

So $\qquad (\cos \theta)^2 + (\sin \theta)^2 \equiv 1$

Hint: Positive powers of trigonometric functions are written without brackets but with the power before the angle: for example $(\cos \theta)^n$ is written as $\cos^n \theta$, if n is positive.

■ For all values of θ, $\cos^2 \theta + \sin^2 \theta \equiv 1$.

Hint: This is sometimes known as Pythagoras' theorem in trigonometry.

Example 1

Simplify the following expressions:

a $\sin^2 3\theta + \cos^2 3\theta$

b $5 - 5\sin^2\theta$

c $\dfrac{\sin 2\theta}{\sqrt{1 - \sin^2 2\theta}}$

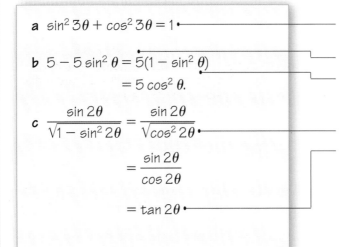

a $\sin^2 3\theta + \cos^2 3\theta = 1$ ———— $\sin^2\theta + \cos^2\theta \equiv 1$, with θ replaced by 3θ.

b $5 - 5\sin^2\theta = 5(1 - \sin^2\theta)$

$\qquad = 5\cos^2\theta.$

———— Always look for factors.

$\sin^2\theta + \cos^2\theta \equiv 1$, so $1 - \sin^2\theta \equiv \cos^2\theta$.

c $\dfrac{\sin 2\theta}{\sqrt{1 - \sin^2 2\theta}} = \dfrac{\sin 2\theta}{\sqrt{\cos^2 2\theta}}$

$\qquad = \dfrac{\sin 2\theta}{\cos 2\theta}$

$\qquad = \tan 2\theta$

———— $\sin^2 2\theta + \cos^2 2\theta \equiv 1$, so $1 - \sin^2 2\theta \equiv \cos^2 2\theta$.

$\tan\theta \equiv \dfrac{\sin\theta}{\cos\theta}$, so $\dfrac{\sin 2\theta}{\cos 2\theta} \equiv \tan 2\theta$.

Example 2

Show that $\dfrac{\cos^4\theta - \sin^4\theta}{\cos^2\theta} \equiv 1 - \tan^2\theta$

When you have to prove an identity like this you may quote the basic identities like '$\sin^2 + \cos^2 \equiv 1$'.

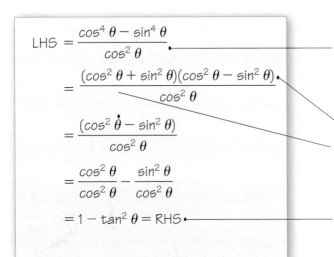

$\text{LHS} = \dfrac{\cos^4\theta - \sin^4\theta}{\cos^2\theta}$

$\qquad = \dfrac{(\cos^2\theta + \sin^2\theta)(\cos^2\theta - \sin^2\theta)}{\cos^2\theta}$

$\qquad = \dfrac{(\cos^2\theta - \sin^2\theta)}{\cos^2\theta}$

$\qquad = \dfrac{\cos^2\theta}{\cos^2\theta} - \dfrac{\sin^2\theta}{\cos^2\theta}$

$\qquad = 1 - \tan^2\theta = \text{RHS}$

Usually the best strategy is to start with the more complicated side (here the left-hand side, LHS) and try to produce the expression on the other side.

The numerator can be factorised as the 'difference of two squares'.
$\sin^2\theta + \cos^2\theta \equiv 1$.

Divide through by $\cos^2\theta$ and note that
$\dfrac{\sin^2\theta}{\cos^2\theta} = \left(\dfrac{\sin\theta}{\cos\theta}\right)^2 = \tan^2\theta.$

Example 3

Given that $\cos \theta = -\frac{3}{5}$ and that θ is reflex, find the value of: **a** $\sin \theta$ and **b** $\tan \theta$.

Method 1

a Since $\sin^2 \theta + \cos^2 \theta \equiv 1$,

$$\sin^2 \theta = 1 - \left(-\frac{3}{5}\right)^2$$

$$= 1 - \frac{9}{25}$$

$$= \frac{16}{25}$$

So $\sin \theta = -\frac{4}{5}$

'θ is reflex' is critical information. Using your calculator to solve $\cos \theta = -\frac{3}{5}$, without consideration of the correct quadrant, would give wrong answers for both $\sin \theta$ and $\tan \theta$.

'θ is reflex' means θ is in 3rd or 4th quadrants, but as $\cos \theta$ is negative, θ must be in the 3rd quadrant. $\sin \theta = \pm\frac{4}{5}$ but in the third quadrant $\sin \theta$ is negative.

b $\tan \theta = \dfrac{\sin \theta}{\cos \theta}$

$$= \frac{\frac{-4}{5}}{\frac{-3}{5}}$$

$$= \frac{4}{3}$$

Method 2

a Use the right-angled triangle with the acute angle ϕ, where $\cos \phi = \dfrac{3}{5}$.

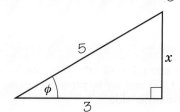

Using Pythagoras' theorem, $x = 4$

$$\text{so } \sin \phi = \frac{4}{5}$$

As $\sin \theta = -\sin \phi$, $\sin \theta = -\dfrac{4}{5}$

b Also from the triangle, $\tan \phi = \dfrac{4}{3}$

As $\tan \theta = +\tan \phi$, $\tan \theta = +\dfrac{4}{3}$

As θ is in the third quadrant you know from the work in Chapter 8, that $\theta = 180 + \phi$

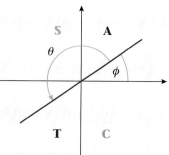

so
$\sin \theta = -\sin \phi$ and $\tan \theta = +\tan \phi$

Hint: Given an angle θ of any size, you can always work with its related acute angle ϕ, and use a right-angled triangle work to find $\sin \phi$, $\cos \phi$ and $\tan \phi$. But remember to consider the quadrant that θ is in when giving $\sin \theta$, $\cos \theta$ and $\tan \theta$.

Example 4

Given that $\sin \alpha = \dfrac{2}{5}$ and that α is obtuse, find the exact value of $\cos \alpha$.

Work with the acute angle ϕ, where $\sin \phi = \dfrac{2}{5}$.

Draw the right-angled triangle and work out the third side.

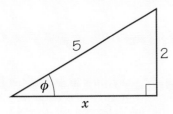

Using Pythagoras' theorem,

$2^2 + x^2 = 5^2,$

so $x^2 = 21 \Rightarrow x = \sqrt{21}$

So $\cos \phi = \dfrac{\sqrt{21}}{5}$

Therefore, $\cos \alpha = -\dfrac{\sqrt{21}}{5}$ as α is in the second quadrant.

'Exact' here means 'Do not use your calculator to find α'.

Alternatively, as in Method 1 of Example 3:
Using $\sin^2 \alpha + \cos^2 \alpha \equiv 1,$

$\cos^2 \alpha = 1 - \dfrac{4}{25} = \dfrac{21}{25}$

As α is obtuse, $\cos \alpha$ is $-$ve,

so $\cos \alpha = -\dfrac{\sqrt{21}}{5}$

Example 5

Given that $p = 3 \cos \theta$, and that $q = 2 \sin \theta$, show that $4p^2 + 9q^2 = 36$.

As $p = 3 \cos \theta$, and $q = 2 \sin \theta$,

so $\cos \theta = \dfrac{p}{3}$ and $\sin \theta = \dfrac{q}{2}$

Using $\sin^2 \theta + \cos^2 \theta \equiv 1,$

$\left(\dfrac{q}{2}\right)^2 + \left(\dfrac{p}{3}\right)^2 = 1$

so $\dfrac{q^2}{4} + \dfrac{p^2}{9} = 1$

$\therefore \quad 9q^2 + 4p^2 = 36$

You need to eliminate θ from the equations. As you can find $\sin \theta$ and $\cos \theta$ in terms of p and q, use the identity $\sin^2 \theta + \cos^2 \theta \equiv 1$.

Multiplying both sides by 36.

Exercise 10A

1 Simplify each of the following expressions:

a $1 - \cos^2 \frac{1}{2}\theta$ **b** $5 \sin^2 3\theta + 5 \cos^2 3\theta$ **c** $\sin^2 A - 1$

d $\dfrac{\sin \theta}{\tan \theta}$ **e** $\dfrac{\sqrt{1 - \cos^2 x}}{\cos x°}$ **f** $\dfrac{\sqrt{1 - \cos^2 3A}}{\sqrt{1 - \sin^2 3A}}$

g $(1 + \sin x)^2 + (1 - \sin x)^2 + 2 \cos^2 x$

h $\sin^4 \theta + \sin^2 \theta \cos^2 \theta$

i $\sin^4 \theta + 2 \sin^2 \theta \cos^2 \theta + \cos^4 \theta$

2 Given that $2 \sin \theta = 3 \cos \theta$, find the value of $\tan \theta$.

3 Given that $\sin x \cos y = 3 \cos x \sin y$, express $\tan x$ in terms of $\tan y$.

4 Express in terms of $\sin \theta$ only:

a $\cos^2 \theta$ **b** $\tan^2 \theta$ **c** $\cos \theta \tan \theta$

d $\dfrac{\cos \theta}{\tan \theta}$ **e** $(\cos \theta - \sin \theta)(\cos \theta + \sin \theta)$

5 Using the identities $\sin^2 A + \cos^2 A \equiv 1$ and/or $\tan A \equiv \dfrac{\sin A}{\cos A}$ $(\cos A \neq 0)$, prove that:

a $(\sin \theta + \cos \theta)^2 \equiv 1 + 2 \sin \theta \cos \theta$ **b** $\dfrac{1}{\cos \theta} - \cos \theta \equiv \sin \theta \tan \theta$

c $\tan x + \dfrac{1}{\tan x} \equiv \dfrac{1}{\sin x \cos x}$ **d** $\cos^2 A - \sin^2 A \equiv 2 \cos^2 A - 1 \equiv 1 - 2 \sin^2 A$

e $(2 \sin \theta - \cos \theta)^2 + (\sin \theta + 2 \cos \theta)^2 \equiv 5$ **f** $2 - (\sin \theta - \cos \theta)^2 \equiv (\sin \theta + \cos \theta)^2$

g $\sin^2 x \cos^2 y - \cos^2 x \sin^2 y = \sin^2 x - \sin^2 y$

6 Find, without using your calculator, the values of:

a $\sin \theta$ and $\cos \theta$, given that $\tan \theta = \frac{5}{12}$ and θ is acute.

b $\sin \theta$ and $\tan \theta$, given that $\cos \theta = -\frac{3}{5}$ and θ is obtuse.

c $\cos \theta$ and $\tan \theta$, given that $\sin \theta = -\frac{7}{25}$ and $270° < \theta < 360°$.

7 Given that $\sin \theta = \frac{2}{3}$ and that θ is obtuse, find the exact value of: **a** $\cos \theta$, **b** $\tan \theta$.

8 Given that $\tan \theta = -\sqrt{3}$ and that θ is reflex, find the exact value of: **a** $\sin \theta$, **b** $\cos \theta$.

9 Given that $\cos \theta = \frac{3}{4}$ and that θ is reflex, find the exact value of: **a** $\sin \theta$, **b** $\tan \theta$.

10 In each of the following, eliminate θ to give an equation relating x and y:

a $x = \sin \theta, y = \cos \theta$

b $x = \sin \theta, y = 2 \cos \theta$

c $x = \sin \theta, y = \cos^2 \theta$

d $x = \sin \theta, y = \tan \theta$

e $x = \sin \theta + \cos \theta, y = \cos \theta - \sin \theta$

10.2 You need to be able to solve simple trigonometrical equations of the form $\sin\theta = k$, $\cos\theta = k$, (where $-1 \le k \le 1$) and $\tan\theta = p$ ($p \in \mathbb{R}$)

You can show solutions to the equation $\sin\theta = \frac{1}{2}$ in the interval $0° \le \theta \le 360°$ by plotting the graphs of $y = \sin\theta$ and $y = \frac{1}{2}$ and seeing where they intersect.

You can find the exact solutions by using what you already know about trigonometrical functions.

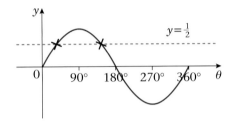

Example 6

Find the solutions of the equation $\sin\theta = \frac{1}{2}$ in the interval $0 \le \theta \le 360°$.

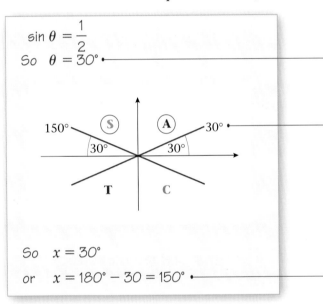

$\sin\theta = \dfrac{1}{2}$

So $\theta = 30°$

Use $\sin^{-1}$ on your calculator to find one solution.

Putting $30°$ in the four positions shown gives the angles $30°$, $150°$, $210°$ and $330°$ but sine is only $+$ve in the 1st and 2nd quadrants.

So $x = 30°$

or $x = 180° - 30 = 150°$

You can check this by putting $\sin 150°$ in your calculator.

Example 7

Solve, in the interval $0 < x \le 360°$, $5\sin x = -2$.

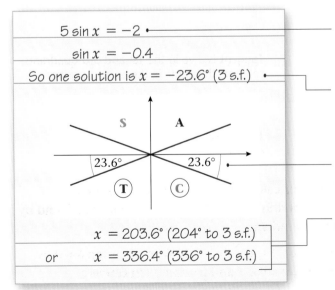

$5\sin x = -2$

$\sin x = -0.4$

So one solution is $x = -23.6°$ (3 s.f.)

First rewrite in the form $\sin x = \ldots$

Note that this calculator solution is not in the given interval.

Sine is $-$ve so you need to look in the 3rd and 4th quadrants for your solutions.

You can now read off the solutions in the given interval.

$x = 203.6°$ ($204°$ to 3 s.f.)

or $x = 336.4°$ ($336°$ to 3 s.f.)

Note that in this case, if $\alpha = \sin^{-1}(-0.4)$ the solutions are $180 - \alpha$ and $360 + \alpha$.

■ A first solution of the equation $\sin x = k$ is your calculator value, $\alpha = \sin^{-1} k$. A second solution is $(180° - \alpha)$, or $(\pi - \alpha)$ if you are working in radians. Other solutions are found by adding or subtracting multiples of 360° or 2π radians.

Example 8

Solve, in the interval $0 < x \leqslant 360°$, $\cos x = \dfrac{\sqrt{3}}{2}$.

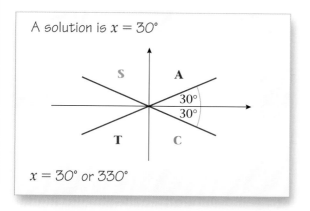

A solution is $x = 30°$

$x = 30°$ or $330°$

The calculator solution of $\cos x = \dfrac{\sqrt{3}}{2}$, is

$x = 30°$, a result you should know.

$\cos x$ is +ve so you need to look in the 1st and 4th quadrants.

Read off the solutions, in $0 < x \leqslant 360°$, from your diagram.

Note that these results are $\alpha°$ and $(360 - \alpha)°$

where $\alpha = \cos^{-1}\left(\dfrac{\sqrt{3}}{2}\right)$.

■ A first solution of the equation $\cos x = k$ is your calculator value $\alpha = \cos^{-1} k$. A second solution is $(360° - \alpha)$, or $(2\pi - \alpha)$ if you are working in radians. Other solutions are found by adding or subtracting multiples of 360° or 2π radians.

Example 9

Find the values of θ, in radians, in the interval $0 < \theta \leqslant 2\pi$, that satisfy the equation $\sin \theta = \sqrt{3} \cos \theta$.

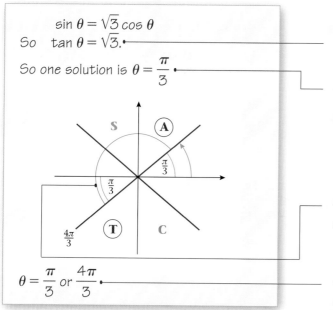

$\sin \theta = \sqrt{3} \cos \theta$

So $\tan \theta = \sqrt{3}$.

So one solution is $\theta = \dfrac{\pi}{3}$

$\theta = \dfrac{\pi}{3}$ or $\dfrac{4\pi}{3}$

As the solutions of $\cos \theta = 0$ do not satisfy the equation you can divide both sides by $\cos \theta$.

This is your calculator answer. (Use radian mode.)

Tangent is +ve in the 1st and 3rd quadrants, so insert the angle in the correct positions.

These results are α and $(\pi + \alpha)$ where $\alpha = \tan^{-1} \sqrt{3}$.

■ A first solution of the equation $\tan x = k$ is your calculator value $\alpha = \tan^{-1} k$. A second solution is $(180° + \alpha)$, or $(\pi + \alpha)$ if you are working in radians. Other solutions are found by adding or subtracting multiples of 360° or 2π radians.

It is easier to find the solutions of $\sin \theta = -1$ or 0 or +1, $\cos \theta = -1$ or 0 or +1 or $\tan \theta = 0$ by considering the corresponding graphs of $y = \sin \theta$, $y = \cos \theta$ and $y = \tan \theta$ respectively.

Exercise 10B

1 Solve the following equations for θ, in the interval $0 < \theta \leqslant 360°$:

a $\sin \theta = -1$

b $\tan \theta = \sqrt{3}$

c $\cos \theta = \frac{1}{2}$

d $\sin \theta = \sin 15°$

e $\cos \theta = -\cos 40°$

f $\tan \theta = -1$

g $\cos \theta = 0$

h $\sin \theta = -0.766$

i $7 \sin \theta = 5$

j $2 \cos \theta = -\sqrt{2}$

k $\sqrt{3} \sin \theta = \cos \theta$

l $\sin \theta + \cos \theta = 0$

m $3 \cos \theta = -2$

n $(\sin \theta - 1)(5 \cos \theta + 3) = 0$

o $\tan \theta = \tan \theta \, (2 + 3 \sin \theta)$

2 Solve the following equations for x, giving your answers to 3 significant figures where appropriate, in the intervals indicated:

a $\sin x° = -\dfrac{\sqrt{3}}{2}, \; -180 \leqslant x \leqslant 540$

b $2 \sin x° = -0.3, \; -180 \leqslant x \leqslant 180$

c $\cos x° = -0.809, \; -180 \leqslant x \leqslant 180$

d $\cos x° = 0.84, \; -360 < x < 0$

e $\tan x° = -\dfrac{\sqrt{3}}{3}, \; 0 \leqslant x \leqslant 720$

f $\tan x° = 2.90, \; 80 \leqslant x \leqslant 440$

3 Solve, in the intervals indicated, the following equations for θ, where θ is measured in radians. Give your answer in terms of π or 2 decimal places.

a $\sin \theta = 0, \; -2\pi < \theta \leqslant 2\pi$

b $\cos \theta = -\frac{1}{2}, \; -2\pi < \theta \leqslant \pi$

c $\sin \theta = \dfrac{1}{\sqrt{2}}, \; -2\pi < \theta \leqslant \pi$

d $\sin \theta = \tan \theta, \; 0 < \theta \leqslant 2\pi$

e $2(1 + \tan \theta) = 1 - 5 \tan \theta, \; -\pi < \theta \leqslant 2\pi$

f $2 \cos \theta = 3 \sin \theta, \; 0 < \theta \leqslant 2\pi$

10.3 You need to be able to solve equations of the form sin $(n\theta + \alpha) = k$, cos $(n\theta + \alpha) = k$, and tan $(n\theta + \alpha) = p$.

You can replace $(n\theta + \alpha)$ by X so that equation reduces to the type you have solved in Section 10.2. You must be careful to ensure that you give all the solutions in the given interval.

Example 10

Solve the equation cos $3\theta° = 0.766$, in the interval $0 \le \theta \le 360$.

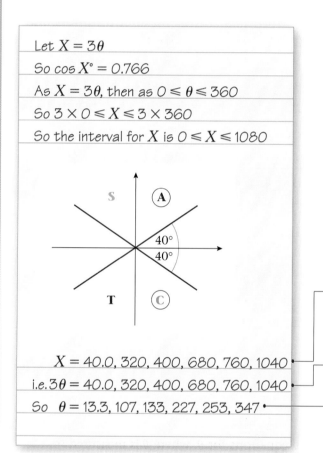

Let $X = 3\theta$

So cos $X° = 0.766$

As $X = 3\theta$, then as $0 \le \theta \le 360$

So $3 \times 0 \le X \le 3 \times 360$

So the interval for X is $0 \le X \le 1080$

$X = 40.0, 320, 400, 680, 760, 1040$

i.e. $3\theta = 40.0, 320, 400, 680, 760, 1040$

So $\theta = 13.3, 107, 133, 227, 253, 347$

Replace 3θ by X and solve.

The value of X from your calculator is 40.0. You need to list all values in the 1st and 4th quadrants for three complete revolutions.

Remember $X = 3\theta$.

Divide by 3.

Always check that all of your solutions are in the given interval.

Example 11

Solve the equation $\sin(2\theta - 35)° = -1$, in the interval $-180 \leqslant \theta \leqslant 180$.

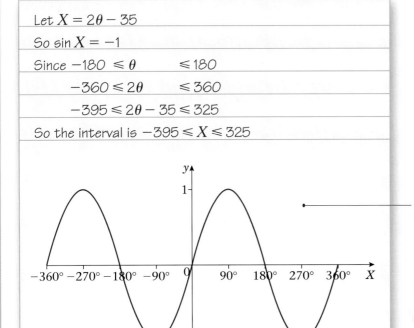

Let $X = 2\theta - 35$

So $\sin X = -1$

Since $-180 \leqslant \theta \leqslant 180$

$\quad\quad -360 \leqslant 2\theta \leqslant 360$

$\quad -395 \leqslant 2\theta - 35 \leqslant 325$

So the interval is $-395 \leqslant X \leqslant 325$

> You can sketch a graph of $\sin X$.

> Refer to the graph for solutions of $\sin X = -1$.

The values of X are $-90, 270$

So $\quad 2\theta - 35 = -90, 270$

$\quad\quad\quad\quad 2\theta = -55, 305$

$\quad\quad\quad\quad\quad \theta = -27.5, 152.5$

Exercise 10C

1 Find the values of θ, in the interval $0 \leqslant \theta \leqslant 360°$, for which:

   a $\sin 4\theta = 0$ **b** $\cos 3\theta = -1$ **c** $\tan 2\theta = 1$ **d** $\cos 2\theta = \frac{1}{2}$

   e $\tan\frac{1}{2}\theta = -\dfrac{1}{\sqrt{3}}$ **f** $\sin(-\theta) = \dfrac{1}{\sqrt{2}}$ **g** $\tan(45° - \theta) = -1$

   h $2\sin(\theta - 20°) = 1$ **i** $\tan(\theta + 75°) = \sqrt{3}$ **j** $\cos(50° + 2\theta) = -1$

2 Solve each of the following equations, in the interval given.
Give your answers to 3 significant figures where appropriate.

   a $\sin(\theta - 10°) = -\dfrac{\sqrt{3}}{2}, \ 0 < \theta \leqslant 360°$ **b** $\cos(70 - x)° = 0.6, \ -180 < x \leqslant 180$

   c $\tan(3x + 25)° = -0.51, \ -90 < x \leqslant 180$ **d** $5\sin 4\theta + 1 = 0, \ -90° \leqslant \theta \leqslant 90°$

3 Solve the following equations for θ, in the intervals indicated. Give your answers in radians.

   a $\sin\left(\theta - \dfrac{\pi}{6}\right) = -\dfrac{1}{\sqrt{2}}, \ -\pi < \theta \leqslant \pi$ **b** $\cos(2\theta + 0.2^c) = -0.2, \ -\dfrac{\pi}{2} \leqslant \theta \leqslant \dfrac{\pi}{2}$

   c $\tan\left(2\theta + \dfrac{\pi}{4}\right) = 1, \ 0 \leqslant \theta \leqslant 2\pi$ **d** $\sin\left(\theta + \dfrac{\pi}{3}\right) = \tan\dfrac{\pi}{6}, \ 0 \leqslant \theta \leqslant 2\pi$

10.4 You need to be able to solve quadratic equations in $\sin\theta$ or $\cos\theta$ or $\tan\theta$. An equation like $\sin^2\theta + 2\sin\theta - 3 = 0$ can be solved in the same way as $x^2 + 2x - 3 = 0$, with $\sin\theta$ replacing x.

Example 12

Solve for θ, in the interval $0 \leqslant \theta \leqslant 360°$, the equations

a $2\cos^2\theta - \cos\theta - 1 = 0$

b $\sin^2(\theta - 30°) = \frac{1}{2}$

a $2\cos^2\theta - \cos\theta - 1 = 0$

So $(2\cos\theta + 1)(\cos\theta - 1) = 0$ •————— Compare with $2x^2 - x - 1 \equiv (2x + 1)(x - 1)$

So $\cos\theta = -\dfrac{1}{2}$ or $\cos\theta = 1$

————— Find one solution using your calculator.

$\cos\theta = -\dfrac{1}{2}$ so $\theta = 120°$

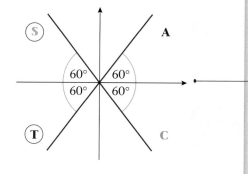

————— 120° makes an angle of 60° with the horizontal. But cosine is $-$ve in the 2nd and 3rd quadrants so $\theta = 120°$ or $\theta = 240°$.

$\theta = 120°$ or $\theta = 240°$

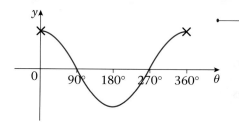

————— Sketch the graph of $y = \cos\theta$.

So $\theta = 0$ or $360°$

So the solutions are $\theta = 0, 120, 240, 360$

b $\sin^2(\theta - 30°) = \dfrac{1}{2}$

So $\sin(\theta - 30°) = \dfrac{1}{\sqrt{2}}$

or $\sin(\theta - 30°) = -\dfrac{1}{\sqrt{2}}$ •————— The solutions of $x^2 = k$ are $x = \pm\sqrt{k}$.

So $\theta - 30° = 45°$ or $\theta - 30° = -45°$ •————— Use your calculator to find one solution for each equation.

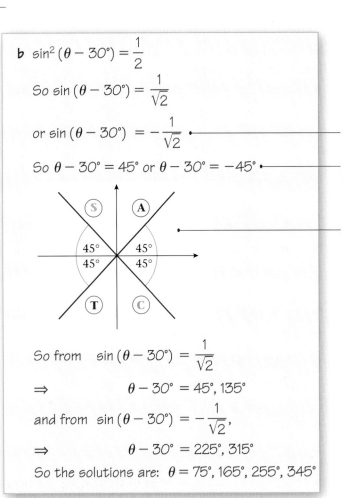

•————— Draw a diagram to find the quadrants where sine is +ve and the quadrants where sine is −ve.

So from $\sin(\theta - 30°) = \dfrac{1}{\sqrt{2}}$

$\Rightarrow \qquad\qquad \theta - 30° = 45°, 135°$

and from $\sin(\theta - 30°) = -\dfrac{1}{\sqrt{2}},$

$\Rightarrow \qquad\qquad \theta - 30° = 225°, 315°$

So the solutions are: $\theta = 75°, 165°, 255°, 345°$

In some equations you may need to use the identity $\sin^2 A + \cos^2 A \equiv 1$ before they are in a form to be solved by factorising (or use of 'the formula').

Example 13

Find the values of x, in the interval $-\pi \leqslant x \leqslant \pi$, satisfying the equation $2\cos^2 x + 9\sin x = 3\sin^2 x$.

$2\cos^2 x + 9\sin x = 3\sin^2 x$ can be written as

$2(1 - \sin^2 x) + 9\sin x = 3\sin^2 x$ •————— As $\sin^2 x + \cos^2 x \equiv 1$, you are able to rewrite $\cos^2 x$ as $(1 - \sin^2 x)$, and so form a quadratic equation in $\sin x$.

which reduces to

$\qquad 5\sin^2 x - 9\sin x - 2 = 0$

So $(5\sin x + 1)(\sin x - 2) = 0$

$\qquad\qquad\qquad \sin x = -\dfrac{1}{5}$ •————— The other equation, $\sin x = 2$, has no solutions.

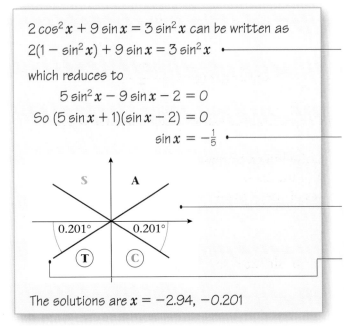

•————— Your calculator value of x, in radian mode, is $x = -0.201$ (3 s.f.). Insert in the quadrant diagram.

————— The smallest angle in the interval, in the 3rd quadrant, is $(-\pi + 0.201) = -2.94$; there are no values of x between 0 and π.

The solutions are $x = -2.94, -0.201$

Exercise 10D

1 Solve for θ, in the interval $0 \leqslant \theta \leqslant 360°$, the following equations.
Give your answers to 3 significant figures where they are not exact.

a $4\cos^2\theta = 1$
b $2\sin^2\theta - 1 = 0$

c $3\sin^2\theta + \sin\theta = 0$
d $\tan^2\theta - 2\tan\theta - 10 = 0$

e $2\cos^2\theta - 5\cos\theta + 2 = 0$
f $\sin^2\theta - 2\sin\theta - 1 = 0$

g $\tan^2 2\theta = 3$
h $4\sin\theta = \tan\theta$

i $\sin\theta + 2\cos^2\theta + 1 = 0$
j $\tan^2(\theta - 45°) = 1$

k $3\sin^2\theta = \sin\theta\cos\theta$
l $4\cos\theta(\cos\theta - 1) = -5\cos\theta$

m $4(\sin^2\theta - \cos\theta) = 3 - 2\cos\theta$
n $2\sin^2\theta = 3(1 - \cos\theta)$

o $4\cos^2\theta - 5\sin\theta - 5 = 0$
p $\cos^2\dfrac{\theta}{2} = 1 + \sin\dfrac{\theta}{2}$

2 Solve for θ, in the interval $-180° \leqslant \theta \leqslant 180°$, the following equations.
Give your answers to 3 significant figures where they are not exact.

a $\sin^2 2\theta = 1$
b $\tan^2\theta = 2\tan\theta$

c $\cos\theta(\cos\theta - 2) = 1$
d $\sin^2(\theta + 10°) = 0.8$

e $\cos^2 3\theta - \cos 3\theta = 2$
f $5\sin^2\theta = 4\cos^2\theta$

g $\tan\theta = \cos\theta$
h $2\sin^2\theta + 3\cos\theta = 1$

3 Solve for x, in the interval $0 \leqslant x \leqslant 2\pi$, the following equations.
Give your answers to 3 significant figures unless they can be written in the form $\dfrac{a}{b}\pi$, where a and b are integers.

a $\tan^2\frac{1}{2}x = 1$
b $2\sin^2\left(x + \dfrac{\pi}{3}\right) = 1$

c $3\tan x = 2\tan^2 x$
d $\sin^2 x + 2\sin x\cos x = 0$

e $6\sin^2 x + \cos x - 4 = 0$
f $\cos^2 x - 6\sin x = 5$

g $2\sin^2 x = 3\sin x\cos x + 2\cos^2 x$

Mixed exercise 10E

1 Given that angle A is obtuse and $\cos A = -\sqrt{\dfrac{7}{11}}$, show that $\tan A = \dfrac{-2\sqrt{7}}{7}$.

2 Given that angle B is reflex and $\tan B = +\dfrac{\sqrt{21}}{2}$, find the exact value of: **a** $\sin B$, **b** $\cos B$.

3 **a** Sketch the graph of $y = \sin(x + 60)°$, in the interval $-360 \leqslant x \leqslant 360$, giving the coordinates of points of intersection with the axes.
b Calculate the values of the x-coordinates of the points in which the line $y = \frac{1}{2}$ intersects the curve.

4 Simplify the following expressions:

a $\cos^4\theta - \sin^4\theta$
b $\sin^2 3\theta - \sin^2 3\theta\cos^2 3\theta$

c $\cos^4\theta + 2\sin^2\theta\cos^2\theta + \sin^4\theta$

5 **a** Given that $2\,(\sin x + 2\cos x) = \sin x + 5\cos x$, find the exact value of $\tan x$.

   b Given that $\sin x \cos y + 3\cos x \sin y = 2\sin x \sin y - 4\cos x \cos y$, express $\tan y$ in terms of $\tan x$.

6 Show that, for all values of θ:

   a $(1 + \sin\theta)^2 + \cos^2\theta = 2(1 + \sin\theta)$ **b** $\cos^4\theta + \sin^2\theta = \sin^4\theta + \cos^2\theta$

7 Without attempting to solving them, state how many solutions the following equations have in the interval $0 \leqslant \theta \leqslant 360°$. Give a brief reason for your answer.

   a $2\sin\theta = 3$ **b** $\sin\theta = -\cos\theta$

   c $2\sin\theta + 3\cos\theta + 6 = 0$ **d** $\tan\theta + \dfrac{1}{\tan\theta} = 0$

8 **a** Factorise $4xy - y^2 + 4x - y$.

   b Solve the equation $4\sin\theta\cos\theta - \cos^2\theta + 4\sin\theta - \cos\theta = 0$, in the interval $0 \leqslant \theta \leqslant 360°$.

9 **a** Express $4\cos 3\theta° - \sin(90 - 3\theta)°$ as a single trigonometric function.

   b Hence solve $4\cos 3\theta° - \sin(90 - 3\theta)° = 2$ in the interval $0 \leqslant \theta \leqslant 360$.
   Give your answers to 3 significant figures.

10 Find, in radians to two decimal places, the value of x in the interval $0 \leqslant x \leqslant 2\pi$, for which $3\sin^2 x + \sin x - 2 = 0$. **E**

11 Given that $2\sin 2\theta = \cos 2\theta$:

   a Show that $\tan 2\theta = 0.5$.

   b Hence find the value of θ, to one decimal place, in the interval $0 \leqslant \theta < 360°$ for which $2\sin 2\theta° = \cos 2\theta°$. **E**

12 Find all the values of θ in the interval $0 \leqslant \theta < 360$ for which:

   a $\cos(\theta + 75)° = 0.5$.

   b $\sin 2\theta° = 0.7$, giving your answers to one decimal place. **E**

13 **a** Find the coordinates of the point where the graph of $y = 2\sin(2x + \tfrac{5}{6}\pi)$ crosses the y-axis.

   b Find the values of x, where $0 \leqslant x \leqslant 2\pi$, for which $y = \sqrt{2}$. **E**

14 Find, giving your answers in terms of π, all values of θ in the interval $0 < \theta < 2\pi$, for which:

   a $\tan\left(\theta + \dfrac{\pi}{3}\right) = 1$ **b** $\sin 2\theta = -\dfrac{\sqrt{3}}{2}$ **E**

15 Find the values of x in the interval $0 < x < 270°$ which satisfy the equation

$$\frac{\cos 2x + 0.5}{1 - \cos 2x} = 2$$

16 Find, to the nearest integer, the values of x in the interval $0 \leqslant x < 180°$ for which
$3 \sin^2 3x - 7 \cos 3x - 5 = 0$.

17 Find, in degrees, the values of θ in the interval $0 \leqslant \theta < 360°$ for which
$$2 \cos^2 \theta - \cos \theta - 1 = \sin^2 \theta$$
Give your answers to 1 decimal place, where appropriate.

18 Consider the function f(x) defined by

$$\mathrm{f}(x) \equiv 3 + 2 \sin (2x + k)°, \ 0 < x < 360$$

where k is a constant and $0 < k < 360$. The curve with equation $y = \mathrm{f}(x)$ passes through the point with coordinates $(15, 3 + \sqrt{3})$.

a Show that $k = 30$ is a possible value for k and find the other possible value of k.

b Given that $k = 30$, solve the equation f(x) = 1.

19 **a** Determine the solutions of the equation
$\cos (2x - 30)° = 0$ for which $0 \leqslant x \leqslant 360$.

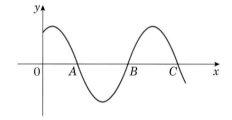

b The diagram shows part of the curve with equation
$y = \cos (px - q)°$, where p and q
are positive constants and $q < 180$. The curve cuts
the x-axis at points A, B and C, as shown.

Given that the coordinates of A and B are (100, 0) and (220, 0) respectively:

i Write down the coordinates of C.

ii Find the value of p and the value of q.

20 The diagram shows part of the curve with equation
$y = \mathrm{f}(x)$, where f(x) = $1 + 2 \sin (px° + q°)$, p and q
being positive constants and $q \leqslant 90$. The curve cuts
the y-axis at the point A and the x-axis at the
points C and D. The point B is a maximum point
on the curve.

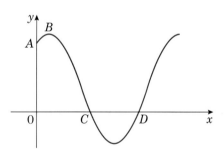

Given that the coordinates of A and C are (0, 2)
and (45, 0) respectively:

a Calculate the value of q.

b Show that $p = 4$.

c Find the coordinates of B and D.

Summary of key points

1   $\tan \theta = \dfrac{\sin \theta}{\cos \theta}$ (providing $\cos \theta \neq 0$, when $\tan \theta$ is not defined)

2   $\sin^2 \theta + \cos^2 \theta = 1$

3   A first solution of the equation $\sin x = k$ is your calculator value, $\alpha = \sin^{-1} k$. A second solution is $(180° - \alpha)$, or $(\pi - \alpha)$ if you are working in radians. Other solutions are found by adding or subtracting multiples of 360° or 2π radians.

4   A first solution of the equation $\cos x = k$ is your calculator value of $\alpha = \cos^{-1} k$. A second solution is $(360° - \alpha)$, or $(2\pi - \alpha)$ if you are working in radians. Other solutions are found by adding or subtracting multiples of 360° or 2π radians.

5   A first solution of the equation $\tan x = k$ is your calculator value $\alpha = \tan^{-1} k$. A second solution is $(180° + \alpha)$, or $(\pi + \alpha)$ if you are working in radians. Other solutions are found by adding or subtracting multiples of 360° or 2π radians.

11 | Integration

This chapter introduces you to definite integration and its application to finding areas.

11.1 You need to be able to integrate simple functions within defined limits. This is called definite integration.

You met indefinite integration in Book C1:

$$\int 3x^2 \, dx = \frac{3x^{2+1}}{3} + C$$
$$= x^3 + C$$

where C is an arbitrary constant.

You can also integrate a function between defined limits, e.g. $x = 1$ and $x = 2$. You write this as

$$\int_1^2 3x^2 \, dx$$

Here is how you work out this definite integral:

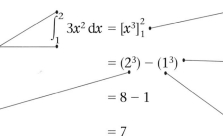

Hint: The limits of the integral are from $x = 1$ to $x = 2$.

Hint: Notice the use of [ ] brackets. This is standard notation.

$$\int_1^2 3x^2 \, dx = [x^3]_1^2$$

Hint: Evaluate the integral at the upper limit.

$$= (2^3) - (1^3)$$

Hint: Notice the use of ( ). This is standard notation for this step.

$$= 8 - 1$$

$$= 7$$

Hint: Evaluate the integral at the lower limit.

There are three stages when you work out a definite integral:

The statement		*After integration* [square brackets]		*The evaluation* (round brackets)
$\int_a^b \ldots dx$	$=$	$[\ldots]_a^b$	$=$	$(\ldots) - (\ldots)$

You should note the notation for evaluating definite integrals and aim to use it when answering questions.

■ **The definite integral is defined as**

$$\int_a^b f'(x) \, dx = [f(x)]_a^b = f(b) - f(a)$$

provided f′ is the derived function of f throughout the interval (a, b).

Example 1

Evaluate the following

a $\displaystyle\int_{1}^{4} (2x - 3x^{\frac{1}{2}} + 1)\,dx$ **b** $\displaystyle\int_{-1}^{0} (x^{\frac{1}{3}} - 1)^2\,dx.$

a $\displaystyle\int_{1}^{4} (2x - 3x^{\frac{1}{2}} + 1)\,dx$

$= \left[x^2 - \dfrac{3x^{\frac{3}{2}}}{\frac{3}{2}} + x \right]_{1}^{4}$

Remember:

$\displaystyle\int_{1}^{4} x^n\,dx = \left[\dfrac{x^{n+1}}{n+1} \right]_{1}^{4}$

$= [x^2 - 2x^{\frac{3}{2}} + x]_{1}^{4}$ ——————— Simplify the terms.

$= (4^2 - 2 \times 2^3 + 4) - (1 - 2 + 1)$ ——— Evaluate expression at $x = 1$.

$= 4 - 0$

$= 4.$ ———————————— Evaluate expression at $x = 4$.

Note that $4^{\frac{3}{2}} = 2^3$.

b $\displaystyle\int_{-1}^{0} (x^{\frac{1}{3}} - 1)^2\,dx$

$= \displaystyle\int_{-1}^{0} (x^{\frac{2}{3}} - 2x^{\frac{1}{3}} + 1)\,dx$ ——— First multiply out the bracket to put the expression in a form ready to be integrated. (See Book C1.)

$= \left[\dfrac{x^{\frac{5}{3}}}{\frac{5}{3}} - 2\dfrac{x^{\frac{4}{3}}}{\frac{4}{3}} + x \right]_{-1}^{0}$ ——— Remember:

$\displaystyle\int_{-1}^{0} x^{\frac{a}{b}}\,dx = \left[\dfrac{x^{\frac{a}{b}+1}}{(\frac{a}{b}+1)} \right]_{-1}^{0}$

$= \left[\dfrac{3}{5}x^{\frac{5}{3}} - \dfrac{6}{4}x^{\frac{4}{3}} + x \right]_{-1}^{0}$

$= (0 + 0 + 0) - \left(-\dfrac{3}{5} - \dfrac{3}{2} - 1 \right)$ ——— Simplify each term.

$= 3\frac{1}{10}$ or 3.1 ——————————— Note that $(-1)^{\frac{4}{3}} = +1$

Exercise 11A

1 Evaluate the following definite integrals:

a $\displaystyle\int_{1}^{2} \left(\dfrac{2}{x^3} + 3x \right) dx$ **b** $\displaystyle\int_{0}^{2} (2x^3 - 4x + 5)\,dx$

c $\displaystyle\int_{4}^{9} \left(\sqrt{x} - \dfrac{6}{x^2} \right) dx$ **d** $\displaystyle\int_{1}^{2} \left(6x - \dfrac{12}{x^4} + 3 \right) dx$

e $\displaystyle\int_{1}^{8} (x^{-\frac{1}{3}} + 2x - 1)\,dx$

2 Evaluate the following definite integrals:

a $\int_1^3 \left(\frac{x^3 + 2x^2}{x}\right) dx$

b $\int_1^4 (\sqrt{x} - 3)^2 \, dx$

c $\int_3^6 \left(x - \frac{3}{x}\right)^2 dx$

d $\int_0^1 x^2\left(\sqrt{x} + \frac{1}{x}\right) dx$

e $\int_1^4 \frac{2 + \sqrt{x}}{x^2} \, dx$

11.2 You need to be able to use definite integration to find areas under curves.

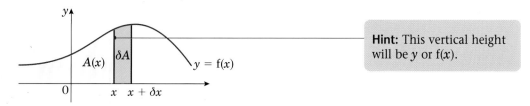

For any curve with equation $y = f(x)$, you can define the area under the curve and to the left of x as a function of x called $A(x)$. As x increases this area $A(x)$ also increases (since x moves further to the right).

If you look at a small increase in x, say δx, then the area increases by an amount $\delta A = A(x + \delta x) - A(x)$.

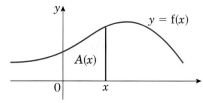

Hint: This vertical height will be y or $f(x)$.

This extra increase in the area δA is approximately rectangular and of magnitude $y\delta x$. (As we make δx smaller any error between the actual area and this will be negligible.)

So we have $\qquad \delta A \approx y\delta x$

or $\qquad \dfrac{\delta A}{\delta x} \approx y$

and if we take the limit $\lim\limits_{\delta x \to 0} \left(\dfrac{\delta A}{\delta x}\right)$ then from Chapter 7 of Book C1 you will see that $\dfrac{dA}{dx} = y$.

Now if you know that $\dfrac{dA}{dx} = y$, then to find A you have to integrate, giving $A = \int y \, dx$.

■ **In particular if you wish to find the area between a curve, the x-axis and the lines $x = a$ and $x = b$ you have**

$$\text{Area} = \int_a^b y \, dx$$

where $y = f(x)$ is the equation of the curve.

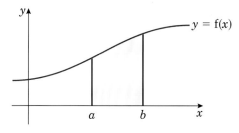

Example 2

Find the area of the region R bounded by the curve with equation $y = (4 - x)(x + 2)$ and the positive x- and y-axes.

When $x = 0$, $y = 8$

When $y = 0$, $x = 4$ or -2

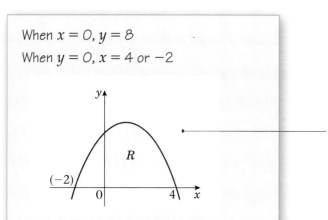

A sketch of the curve will often help in this type of question. (See Chapter 4 of Book C1.)

The area of R is given by

$$A = \int_0^4 (4 - x)(x + 2)\, dx$$

So
$$A = \int_0^4 (8 + 2x - x^2)\, dx$$

Multiply out the brackets.

$$A = \left[8x + x^2 - \frac{x^3}{3} \right]_0^4$$

Integrate.

$$A = \left(32 + 16 - \frac{64}{3} \right) - (0)$$

Use limits of 4 and 0.

So the area is $26\frac{2}{3}$

Example 3

The region R is enclosed by the curve with equation $y = x^2 + \dfrac{4}{x^2}$; $x > 0$, the x-axis and the lines $x = 1$ and $x = 3$. Find the area of R.

This curve is not one you would be expected to sketch but the limits of the integral are simply $x = 1$ and $x = 3$, so you can write down an expression for the area without referring to a sketch.

$$\text{Area} = \int_1^3 \left(x^2 + \frac{4}{x^2} \right) dx = \int_1^3 (x^2 + 4x^{-2})\, dx$$

$$= \left[\frac{x^3}{3} - 4x^{-1} \right]_1^3$$

Write the expression in a form suitable for integrating.

$$= \left(9 - \frac{4}{3} \right) - \left(\frac{1}{3} - 4 \right)$$

Now integrate.

$$= 13 - \frac{5}{3} = 11\frac{1}{3}$$

You may be able to use a graphical calculator to check your answer, but you must show your working.

Exercise 11B

1 Find the area between the curve with equation $y = f(x)$, the x-axis and the lines $x = a$ and $x = b$ in each of the following cases:

a $f(x) = 3x^2 - 2x + 2;$ $a = 0, b = 2$

b $f(x) = x^3 + 4x;$ $a = 1, b = 2$

c $f(x) = \sqrt{x} + 2x;$ $a = 1, b = 4$

d $f(x) = 7 + 2x - x^2;$ $a = -1, b = 2$

e $f(x) = \dfrac{8}{x^3} + \sqrt{x};$ $a = 1, b = 4$

2 The sketch shows part of the curve with equation $y = x(x^2 - 4)$.
Find the area of the shaded region.

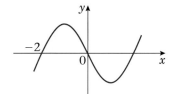

3 The diagram shows a sketch of the curve with equation $y = 3x + \dfrac{6}{x^2} - 5, x > 0$.

The region R is bounded by the curve, the x-axis and the lines $x = 1$ and $x = 3$.
Find the area of R.

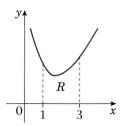

4 Find the area of the finite region between the curve with equation $y = (3 - x)(1 + x)$ and the x-axis.

5 Find the area of the finite region between the curve with equation $y = x(x - 4)^2$ and the x-axis.

6 Find the area of the finite region between the curve with equation $y = x^2(2 - x)$ and the x-axis.

11.3 You need to be able to work out areas of curves under the x-axis.

In the examples so far the area that you were calculating was above the x-axis. If the area between a curve and the x-axis lies below the x-axis, then $\int y \, dx$ will give a negative answer.

Example 4

Find the area of the finite region bounded by the curve $y = x(x - 3)$ and the x-axis.

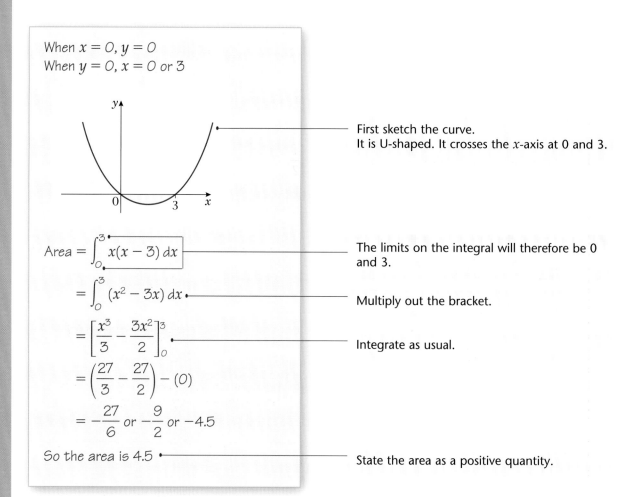

When $x = 0$, $y = 0$

When $y = 0$, $x = 0$ or 3

First sketch the curve.
It is U-shaped. It crosses the x-axis at 0 and 3.

$$\text{Area} = \int_0^3 x(x - 3)\,dx$$

The limits on the integral will therefore be 0 and 3.

$$= \int_0^3 (x^2 - 3x)\,dx$$

Multiply out the bracket.

$$= \left[\frac{x^3}{3} - \frac{3x^2}{2}\right]_0^3$$

Integrate as usual.

$$= \left(\frac{27}{3} - \frac{27}{2}\right) - (0)$$

$$= -\frac{27}{6} \text{ or } -\frac{9}{2} \text{ or } -4.5$$

So the area is 4.5

State the area as a positive quantity.

The following example shows that great care must be taken if you are trying to find an area which straddles the x-axis such as the shaded region below, bounded by the curve with equation $y = (x + 1)(x - 1)x = x^3 - x$.

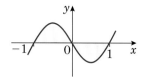

Notice that:

$$\int_{-1}^{1} (x^3 - x)\,dx = \left[\frac{x^4}{4} - \frac{x^2}{2}\right]_{-1}^{1}$$

$$= \left(\frac{1}{4} - \frac{1}{2}\right) - \left(\frac{1}{4} - \frac{1}{2}\right)$$

$$= 0.$$

This is because:

$$\int_0^1 (x^3 - x)\,dx = \left[\frac{x^4}{4} - \frac{x^2}{2}\right]_0^1$$

$$= \left(\frac{1}{4} - \frac{1}{2}\right) - (0) = -\frac{1}{4}$$

and

$$\int_{-1}^0 (x^3 - x)\,dx = \left[\frac{x^4}{4} - \frac{x^2}{2}\right]_{-1}^0$$

$$= -\left(\frac{1}{4} - \frac{1}{2}\right) = \frac{1}{4}$$

So the area of the shaded region is actually $\dfrac{1}{4} + \dfrac{1}{4} = \dfrac{1}{2}$.

For examples of this type you need to draw a sketch, unless one is given in the question.

Example 5

Sketch the curve with equation $y = x(x - 1)(x + 3)$ and find the area of the finite region bounded by the curve and the x-axis.

When $x = 0$, $y = 0$

When $y = 0$, $x = 0$, 1 or -3

Find out where the curve intercepts the axes.

$x \rightarrow \infty$, $y \rightarrow \infty$

$x \rightarrow -\infty$, $y \rightarrow -\infty$

Find out what happens to y when x is large and positive or large and negative.

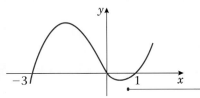

Since the area between $x = 0$ and 1 is below the axis the integral between these points will give a negative answer.

The area is given by $\displaystyle\int_{-3}^0 y\,dx + -\int_0^1 y\,dx$

Now $\displaystyle\int y\,dx = \int (x^3 + 2x^2 - 3x)\,dx$

Multiply out the brackets.

$$= \left[\frac{x^4}{4} + \frac{2x^3}{3} - \frac{3x^2}{2}\right]$$

So $\displaystyle\int_{-3}^0 y\,dx = (0) - \left(\frac{81}{4} - \frac{2}{3} \times 27 - \frac{3}{2} \times 9\right)$

$$= 11.25$$

and $\displaystyle\int_0^1 y\,dx = \left(\frac{1}{4} + \frac{2}{3} - \frac{3}{2}\right) - (0)$

$$= -\frac{7}{12}$$

So the area required is $11.25 + \dfrac{7}{12} = 11\frac{5}{6}$

Exercise 11C

Sketch the following and find the area of the finite region or regions bounded by the curves and the x-axis:

1 $y = x(x + 2)$

2 $y = (x + 1)(x - 4)$

3 $y = (x + 3)x(x - 3)$

4 $y = x^2(x - 2)$

5 $y = x(x - 2)(x - 5)$

11.4 You need to be able to work out the area between a curve and a straight line.

Sometimes you may wish to find an area between a curve and a line. (The method also applies to finding the area between two curves, but this is not required in C2.)

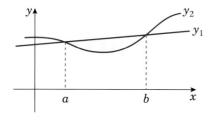

You find the area of the shaded region by calculating $\int_a^b (y_1 - y_2)\, dx$. This is because $\int_a^b y_1\, dx$ gives the area below the line (or curve) with equation y_1, and $\int_a^b y_2\, dx$ gives the area below y_2. So the shaded region is simply $\int_a^b y_1\, dx - \int_a^b y_2\, dx = \int_a^b (y_1 - y_2)\, dx$.

■ **The area between a line (equation y_1) and a curve (equation y_2) is given by**

$$\text{Area} = \int_a^b (y_1 - y_2)\, dx$$

Example 6

The diagram shows a sketch of part of the curve with equation $y = x(4 - x)$ and the line with equation $y = x$.

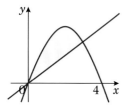

Find the area of the region bounded by the curve and the line.

Method 1

$$x = x(4 - x)$$
$$x = 4x - x^2$$
$$x^2 - 3x = 0$$
$$x(x - 3) = 0$$

So $\quad x = 0 \text{ or } 3$

So the line cuts the curve at $(0, 0)$ and $(3, 3)$

The area is given by $\int_0^3 [x(4 - x) - x]\, dx$

$$\text{Area} = \int_0^3 (3x - x^2)\, dx$$
$$= \left[\frac{3}{2}x^2 - \frac{x^3}{3}\right]_0^3$$
$$= \left(\frac{27}{2} - 9\right) - (0) = 4.5$$

First find where the line and the curve cross.

Find y-coordinates by substituting back in one of the equations. The line $y = x$ is the simplest.

Use the formula with 'curve − line' since the curve is above the line.

Simplify the expression to be integrated.

Method 2

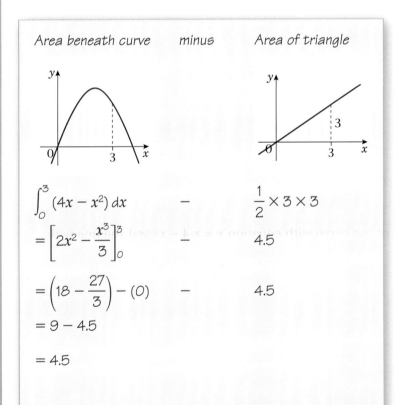

Area beneath curve $\quad$ minus $\quad$ Area of triangle

$$\int_0^3 (4x - x^2)\, dx \qquad - \qquad \frac{1}{2} \times 3 \times 3$$
$$= \left[2x^2 - \frac{x^3}{3}\right]_0^3 \qquad - \qquad 4.5$$
$$= \left(18 - \frac{27}{3}\right) - (0) \qquad - \qquad 4.5$$
$$= 9 - 4.5$$
$$= 4.5$$

You should notice that you could have found this area by first finding the area beneath the curve between $x = 0$ and $x = 3$, and then subtracting the area of a triangle.

The $\displaystyle\int_a^b (y_1 - y_2)\, dx$ formula can be applied even if part of the region is below the x-axis. Consider the following:

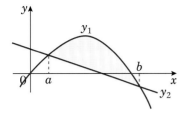

If both the curve and the line are translated upwards by $+k$, where k is sufficiently large to ensure that the required area is totally above the x-axis, the diagram will look like this:

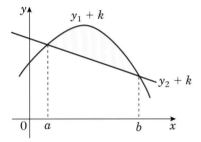

You should notice that since the translation is in the y-direction only, then the x-coordinates of the points of intersection are unchanged and so the limits of the integral will remain the same.

So the area in this case is given by $\displaystyle\int_a^b [y_1 + k - (y_2 + k)]\, dx$

$$= \int_a^b (y_1 - y_2)\, dx$$

Notice that the value of k does not appear in the final formula so you can always use this approach for questions of this type.

Sometimes you will need to add or subtract an area found by integration to the area of a triangle, trapezium or other similar shape as the following example shows.

Example 7

The diagram shows a sketch of the curve with equation $y = x(x - 3)$ and the line with equation $y = 2x$.

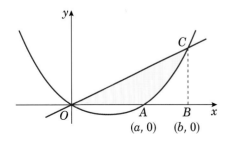

Find the area of the shaded region OAC.

The required area is given by:

$$\text{Area of triangle } OBC - \int_a^b x(x-3)\,dx$$

The curve cuts the x-axis at $x = 3$

(and $x = 0$) so $a = 3$

The curve meets the line $y = 2x$ when

$2x = x(x-3)$ •————————————————— Line = curve.

So $\qquad 0 = x^2 - 5x$ •—————————————

$\qquad\qquad 0 = x(x-5)$ └——— Simplify the equation.

$\qquad\qquad x = 0$ or 5, so $b = 5$

The point C is $(5, 10)$ •————————— $y = 2 \times 5 = 10$, substituting

Area of triangle $OBC = \frac{1}{2} \times 5 \times 10 = 25$ $\qquad x = 5$ into the equation of the line.

Area between curve, x-axis and the line $x = 5$ is

$$\int_3^5 x(x-3)\,dx = \int_3^5 (x^2 - 3x)\,dx$$

$$= \left[\frac{x^3}{3} - \frac{3x^2}{2}\right]_3^5$$

$$= \left(\frac{125}{3} - \frac{75}{2}\right) - \left(\frac{27}{3} - \frac{27}{2}\right)$$

$$= \left(\frac{25}{6}\right) - \left(-\frac{27}{6}\right)$$

$$= \frac{52}{6} \text{ or } \frac{26}{3}$$

Shaded region is therefore $= 25 - \dfrac{26}{3} = \dfrac{49}{3}$ or $16\frac{1}{3}$

Exercise 11D

1 The diagram shows part of the curve with equation $y = x^2 + 2$ and the line with equation $y = 6$. The line cuts the curve at the points A and B.

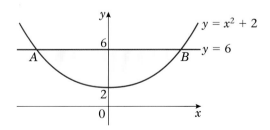

a Find the coordinates of the points A and B.

b Find the area of the finite region bounded by AB and the curve.

2 The diagram shows the finite region, R, bounded by the curve with equation $y = 4x - x^2$ and the line $y = 3$. The line cuts the curve at the points A and B.

a Find the coordinates of the points A and B.

b Find the area of R.

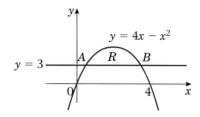

3 The diagram shows a sketch of part of the curve with equation $y = 9 - 3x - 5x^2 - x^3$ and the line with equation $y = 4 - 4x$. The line cuts the curve at the points A $(-1, 8)$ and B $(1, 0)$.

Find the area of the shaded region between AB and the curve.

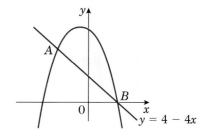

4 Find the area of the finite region bounded by the curve with equation $y = (1 - x)(x + 3)$ and the line $y = x + 3$.

5 The diagram shows the finite region, R, bounded by the curve with equation $y = x(4 + x)$, the line with equation $y = 12$ and the y-axis.

a Find the coordinate of the point A where the line meets the curve.

b Find the area of R.

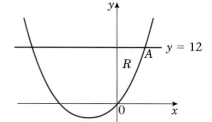

6 The diagram shows a sketch of part of the curve with equation $y = x^2 + 1$ and the line with equation $y = 7 - x$. The finite region R_1 is bounded by the line and the curve. The finite region R_2 is below the curve and the line and is bounded by the positive x- and y-axes as shown in the diagram.

a Find the area of R_1.

b Find the area of R_2.

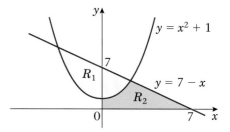

7 The curve C has equation $y = x^{\frac{2}{3}} - \dfrac{2}{x^{\frac{1}{3}}} + 1$.

a Verify that C crosses the x-axis at the point $(1, 0)$.

b Show that the point A $(8, 4)$ also lies on C.

c The point B is $(4, 0)$. Find the equation of the line through AB.
The finite region R is bounded by C, AB and the positive x-axis.

d Find the area of R.

8 The diagram shows part of a sketch of the curve with equation $y = \dfrac{2}{x^2} + x$. The points A and B have x-coordinates $\frac{1}{2}$ and 2 respectively.

Find the area of the finite region between AB and the curve.

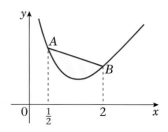

9 The diagram shows part of the curve with equation
$y = 3\sqrt{x} - \sqrt{x^3} + 4$ and the line with equation $y = 4 - \frac{1}{2}x$.

   a Verify that the line and the curve cross at the point A (4, 2).

   b Find the area of the finite region bounded by the curve and the line.

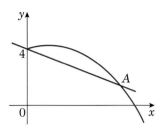
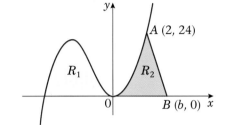

10 The sketch shows part of the curve with equation
$y = x^2(x + 4)$. The finite region R_1 is bounded by the
curve and the negative x-axis. The finite region R_2 is
bounded by the curve, the positive x-axis and AB,
where A (2, 24) and B (b, 0).

The area of R_1 = the area of R_2.
   a Find the area of R_1.
   b Find the value of b.

11.5 Sometimes you may want to find the area beneath a curve but you may not be able to integrate the equation. You can find an approximation to the area using the trapezium rule.

Consider the curve $y = f(x)$:

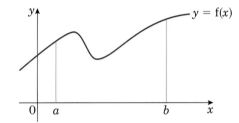

To find the area given by $\int_a^b y \, dx$, we divide the area up
into n equal strips. Each strip will be of width h, so $h = \dfrac{b - a}{n}$.

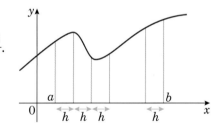

Next we calculate the value of y for each value of x that forms a
boundary of one of the strips. So we find y for $x = a$, $x = a + h$,
$x = a + 2h$, $x = a + 3h$ and so on up to $x = b$. We can label these
values $y_0, y_1, y_2, y_3, \ldots, y_n$.

Hint: Notice that for n
strips there will be $n + 1$
values of x and $n + 1$
values of y.

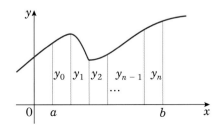

Finally we join adjacent points to form n trapeziums and approximate the original area by the sum of the areas of these n trapeziums.

You may recall from GCSE maths that the area of a trapezium like this:

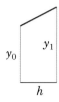

is given by $\frac{1}{2}(y_0 + y_1)h$. The required area under the curve is therefore given by:

$$\int_a^b y\,dx \approx \tfrac{1}{2}h(y_0 + y_1) + \tfrac{1}{2}h(y_1 + y_2) + \ldots + \tfrac{1}{2}h(y_{n-1} + y_n)$$

Factorising gives:

$$\int_a^b y\,dx \approx \tfrac{1}{2}h(y_0 + y_1 + y_1 + y_2 + y_2 \ldots + y_{n-1} + y_{n-1} + y_n)$$

or $$\int_a^b y\,dx \approx \tfrac{1}{2}h[y_0 + 2(y_1 + y_2 \ldots + y_{n-1}) + y_n]$$

This formula is given in the Edexcel formula booklet but you will need to know how to use it.

- **The trapezium rule:**
$$\int_a^b y\,dx \approx \tfrac{1}{2}h[y_0 + 2(y_1 + y_2 \ldots + y_{n-1}) + y_n]$$
  where $h = \dfrac{b-a}{n}$ and $y_i = f(a + ih)$.

Hint: You do not need to remember how to develop this result.

Example 8

Use the trapezium rule with
a 4 strips **b** 8 strips
to estimate the area under the curve with equation $y = \sqrt{(2x + 3)}$ between the lines $x = 0$ and $x = 2$.

a Each strip will have width $\dfrac{2-0}{4} = 0.5$.

x	0	0.5	1	1.5	2
$y = \sqrt{(2x+3)}$	1.732	2	2.236	2.449	2.646

So   area $\approx \dfrac{1}{2} \times 0.5 \times [1.732$

$+ 2(2 + 2.236 + 2.449) + 2.646]$

$= \dfrac{1}{2} \times 0.5 \times [17.748]$

$= 4.437$ or 4.44

First work out the value of y at the boundaries of each of your strips.

It is sometimes helpful to put your working in a table.

b Each strip will be of width $\dfrac{2-0}{8} = 0.25$.

x	0	0.25	0.5	0.75	1	1.25	1.5	1.75	2
y	1.732	1.871	2	2.121	2.236	2.345	2.449	2.550	2.646

So area $\approx \dfrac{1}{2} \times 0.25 \times [1.732 + 2(1.871 + 2 + 2.121 + 2.236 + 2.345$

$$+\ 2.449 + 2.550) + 2.646]$$

$$= \dfrac{1}{2} \times 0.25 \times [35.522]$$

Values of y.

$$= 4.440\ 25 \text{ or } 4.44 \text{ (2 d.p.)}$$

The actual area in this case is 4.441 368 ... and you can see (if you look at the calculations to 3 d.p.) in the above example that increasing the number of strips (or reducing their width) should improve the accuracy of the approximation.

A sketch of $y = \sqrt{(2x + 3)}$ looks like this:

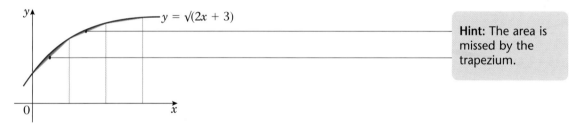

Hint: The area is missed by the trapezium.

You can see that the trapezium rule will always underestimate the area since the curve bends 'outwards'.

In Section 11.2 we mentioned how graphical calculators can be used to evaluate definite integrals. Calculators usually use a slightly different method from the trapezium rule to carry out these calculations and they will generally be more accurate. So, although the calculator can provide a useful check, you should remember that the trapezium rule is being used to *estimate* the value and you should not expect this estimate to be the same as the answer from a graphical calculator.

Exercise | 11E

1 Copy and complete the table below and use the trapezium rule to estimate $\displaystyle\int_1^3 \dfrac{1}{x^2 + 1}\,\mathrm{d}x$:

x	1	1.5	2	2.5	3
$y = \dfrac{1}{x^2 + 1}$	0.5	0.308		0.138	

2 Use the table below to estimate $\displaystyle\int_1^{2.5} \sqrt{(2x - 1)}\,\mathrm{d}x$ with the trapezium rule:

x	1	1.25	1.5	1.75	2	2.25	2.5
$y = \sqrt{(2x - 1)}$	1	1.225	1.414	1.581	1.732	1.871	2

3 Copy and complete the table below and use it, together with the trapezium rule, to estimate $\int_0^2 \sqrt{(x^3 + 1)}\, dx$:

x	0	0.5	1	1.5	2
$y = \sqrt{(x^3 + 1)}$	1	1.061	1.414		

4 **a** Use the trapezium rule with 8 strips to estimate $\int_0^2 2^x\, dx$.

  b With reference to a sketch of $y = 2^x$ explain whether your answer in part **a** is an underestimate or an overestimate of $\int_0^2 2^x\, dx$.

5 Use the trapezium rule with 6 strips to estimate $\int_0^3 \dfrac{1}{\sqrt{(x^2 + 1)}}\, dx$.

6 The diagram shows a sketch of part of the curve with equation $y = \dfrac{1}{x + 2}$, $x > -2$.

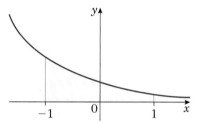

  a Copy and complete the table below and use the trapezium rule to estimate the area bounded by the curve, the x-axis and the lines $x = -1$ and $x = 1$.

x	-1	-0.6	-0.2	0.2	0.6	1
$y = \dfrac{1}{x + 2}$	1	0.714			0.385	0.333

  b State, with a reason, whether your answer in part **a** is an overestimate or an underestimate.

7 **a** Sketch the curve with equation $y = x^3 + 1$, for $-2 < x < 2$.

  b Use the trapezium rule with 4 strips to estimate the value of $\int_{-1}^1 (x^3 + 1)\, dx$.

  c Use integration to find the exact value of $\int_{-1}^1 (x^3 + 1)\, dx$.

  d Comment on your answers to parts **b** and **c**.

8 Use the trapezium rule with 4 strips to estimate $\int_0^2 \sqrt{(3^x - 1)}\, dx$.

9 The sketch shows part of the curve with equation $y = \dfrac{x}{x + 1}$, $x \geqslant 0$.

a Use the trapezium rule with 6 strips to estimate $\int_0^3 \dfrac{x}{x+1}\,dx$.

b With reference to the sketch state, with a reason, whether the answer in part **a** is an overestimate or an underestimate.

10 a Use the trapezium rule with n strips to estimate $\int_0^2 \sqrt{x}\,dx$ in the cases **i** $n = 4$ **ii** $n = 6$.

b Compare your answers from part **a** with the exact value of the integral and calculate the percentage error in each case.

Mixed exercise 11F

1 The diagram shows the curve with equation $y = 5 + 2x - x^2$ and the line with equation $y = 2$. The curve and the line intersect at the points A and B.

a Find the x-coordinates of A and B.

b The shaded region R is bounded by the curve and the line. Find the area of R.

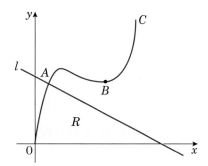

E

2 The diagram shows part of the curve C with equation $y = x^3 - 9x^2 + px$, where p is a constant. The line l has equation $y + 2x = q$, where q is a constant. The point A is the intersection of C and l, and C has a minimum at the point B. The x-coordinates of A and B are 1 and 4 respectively.

a Show that $p = 24$ and calculate the value of q.

b The shaded region R is bounded by C, l and the x-axis. Using calculus, showing all the steps in your working and using the values of p and q found in part **a**, find the area of R.

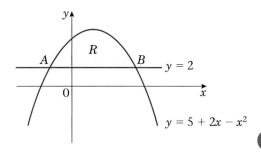

E

3 The diagram shows part of the curve C with equation $y = f(x)$, where $f(x) = 16x^{-\frac{1}{2}} + x^{\frac{3}{2}}$, $x > 0$.

a Use calculus to find the x-coordinate of the minimum point of C, giving your answer in the form $k\sqrt{3}$, where k is an exact fraction.

The shaded region shown in the diagram is bounded by C, the x-axis and the lines with equations $x = 1$ and $x = 2$.

b Using integration and showing all your working, find the area of the shaded region, giving your answer in the form $a + b\sqrt{2}$, where a and b are exact fractions.

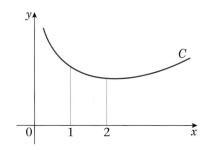

E

4 **a** Find $\int (x^{\frac{1}{2}} - 4)(x^{-\frac{1}{2}} - 1)\, dx$.

b Use your answer to part **a** to evaluate

$$\int_1^4 (x^{\frac{1}{2}} - 4)(x^{-\frac{1}{2}} - 1)\, dx.$$

giving your answer as an exact fraction.　**E**

5 The diagram shows part of the curve with equation $y = x^3 - 6x^2 + 9x$. The curve touches the x-axis at A and has a maximum turning point at B.

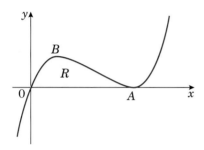

a Show that the equation of the curve may be written as $y = x(x-3)^2$, and hence write down the coordinates of A.

b Find the coordinates of B.

c The shaded region R is bounded by the curve and the x-axis. Find the area of R.　**E**

6 Given that $y^{\frac{1}{2}} = x^{\frac{1}{3}} + 3$:

a Show that $y = x^{\frac{2}{3}} + Ax^{\frac{1}{3}} + B$, where A and B are constants to be found.

b Hence find $\int y\, dx$.

c Using your answer from part **b**, determine the exact value of $\int_1^8 y\, dx$.　**E**

7 Considering the function $y = 3x^{\frac{1}{2}} - 4x^{-\frac{1}{2}}$, $x > 0$:

a Find $\dfrac{dy}{dx}$.

b Find $\int y\, dx$.

c Hence show that $\int_1^3 y\, dx = A + B\sqrt{3}$, where A and B are integers to be found.　**E**

8 The diagram shows a sketch of the curve with equation $y = 2x - x^2$ and the line ON which is the normal to the curve at the origin O.

a Find an equation of ON.

b Show that the x-coordinate of the point N is $2\frac{1}{2}$ and determine its y-coordinate.

c The shaded region shown is bounded by the curve and the line ON. Without using a calculator, determine the area of the shaded region.　**E**

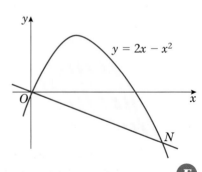

9 The diagram shows a sketch of the curve with equation $y = 12x^{\frac{1}{2}} - x^{\frac{3}{2}}$ for $0 \leqslant x \leqslant 12$.

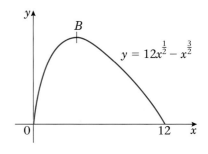

a Show that $\dfrac{dy}{dx} = \dfrac{3}{2}x^{-\frac{1}{2}}(4 - x)$.

b At the point B on the curve the tangent to the curve is parallel to the x-axis. Find the coordinates of the point B.

c Find, to 3 significant figures, the area of the finite region bounded by the curve and the x-axis.

10 The diagram shows the curve C with equation $y = x(8 - x)$ and the line with equation $y = 12$ which meet at the points L and M.

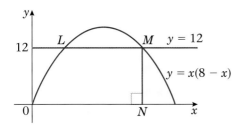

a Determine the coordinates of the point M.

b Given that N is the foot of the perpendicular from M on to the x-axis, calculate the area of the shaded region which is bounded by NM, the curve C and the x-axis.

11 The diagram shows the line $y = x - 1$ meeting the curve with equation $y = (x - 1)(x - 5)$ at A and C. The curve meets the x-axis at A and B.

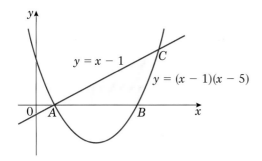

a Write down the coordinates of A and B and find the coordinates of C.

b Find the area of the shaded region bounded by the line, the curve and the x-axis.

12 A and B are two points which lie on the curve C, with equation $y = -x^2 + 5x + 6$. The diagram shows C and the line l passing through A and B.

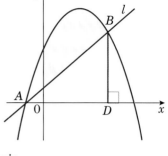

a Calculate the gradient of C at the point where $x = 2$. The line l passes through the point with coordinates $(2, 3)$ and is parallel to the tangent to C at the point where $x = 2$.

b Find an equation of l.

c Find the coordinates of A and B.

The point D is the foot of the perpendicular from B on to the x-axis.

d Find the area of the region bounded by C, the x-axis, the y-axis and BD.

e Hence find the area of the shaded region. **E**

13 The diagram shows part of the curve with equation $y = p + 10x - x^2$, where p is a constant, and part of the line l with equation $y = qx + 25$, where q is a constant. The line l cuts the curve at the points A and B. The x-coordinates of A and B are 4 and 8 respectively. The line through A parallel to the x-axis intersects the curve again at the point C.

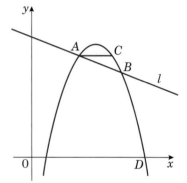

a Show that $p = -7$ and calculate the value of q.

b Calculate the coordinates of C.

c The shaded region in the diagram is bounded by the curve and the line AC. Using algebraic integration and showing all your working, calculate the area of the shaded region. **E**

Summary of key points

1 The definite integral $\int_a^b f'(x)\,dx = f(b) - f(a)$.

2 The area beneath the curve with equation $y = f(x)$ and between the lines $x = a$ and $x = b$ is

$$\text{Area} = \int_a^b f(x)\,dx$$

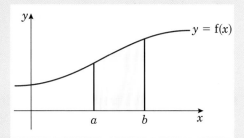

3 The area between a line (equation y_1) and a curve (equation y_2) is given by

$$\text{Area} = \int_a^b (y_1 - y_2)\,dx$$

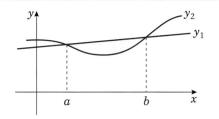

4 **Trapezium rule (in the formula booklet):**

$$\int_a^b y\,dx \approx \tfrac{1}{2}h[y_0 + 2(y_1 + y_2 \ldots + y_{n-1}) + y_n]$$

where $h = \dfrac{b-a}{n}$ and $y_i = f(a + ih)$.

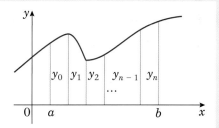

Examination style paper (C2)

(Marks are shown in brackets.)

1 The sector AOB is removed from a circle of radius 5 cm.
The $\angle AOB$ is 1.4 radians and $OA = OB$.

    a Find the perimeter of the sector AOB.   (3)

    b Find the area of sector AOB.   (2)

2 Given that $\log_2 x = p$:

    a Find $\log_2 (8x^2)$ in terms of p.   (4)

    b Given also that $p = 5$, find the value of x.   (2)

3 **a** Find the value of the constant a so that $(x - 3)$ is a factor of $x^3 - ax - 6$.   (3)

    b Using this value of a, factorise $x^3 - ax - 6$ completely.   (4)

4 **a** Find the coefficient of x^{11} and the coefficient of x^{12} in the binomial expansion of $(2 + x)^{15}$.   (4)

    The coefficient of x^{11} and the coefficient of x^{12} in the binomial expansion of $(2 + kx)^{15}$ are equal.

    b Find the value of the constant k.   (3)

5 **a** Prove that:
$$\frac{\cos^2 \theta}{\sin \theta + \sin^2 \theta} \equiv \frac{1 - \sin \theta}{\sin \theta}, \ 0 < \theta < 180°,$$
    (4)

    b Hence, or otherwise, solve the following equation for $0 < \theta < 180°$:
$$\frac{\cos^2 \theta}{\sin \theta + \sin^2 \theta} = 2$$

    Give your answers to the nearest degree.   (4)

6 **a** Show that the centre of the circle with equation $x^2 + y^2 = 6x + 8y$ is (3, 4) and find the radius of the circle.   (5)

    b Find the exact length of the tangents from the point (10, 0) to the circle.   (4)

7 A father promises his daughter an eternal gift on her birthday. On day 1 she receives £75 and each following day she receives $\frac{2}{3}$ of the amount given to her the day before. The father promises that this will go on for ever.

    a Show that after 2 days the daughter will have received £125.   (2)

    b Find how much money the father should set aside to ensure that he can cover the cost of the gift.   (3)

    After k days the total amount of money that the daughter will have received exceeds £200.

    c Find the smallest value of k.   (5)

8 Given $I = \int_{1}^{3} \left(\dfrac{1}{x^2} + 3\sqrt{x} \right) dx$:

  a Use the trapezium rule with the table below to estimate I to 3 significant figures.   (4)

x	1	1.5	2	2.5	3
y	4	4.119	4.493	4.903	5.307

  b Find the exact value of I.   (4)

  c Calculate, to 1 significant figure, the percentage error incurred by using the trapezium rule as in part **a** to estimate I.   (2)

9 The curve C has equation $y = 6x^{\frac{7}{3}} - 7x^2 + 4$.

  a Find $\dfrac{dy}{dx}$.   (2)

  b Find $\dfrac{d^2y}{dx^2}$.   (2)

  c Use your answers to parts **a** and **b** to find the coordinates of the stationary points on C and determine their nature.   (9)

Formulae you need to remember

These are the formulae that you need to remember for your exams. They will not be included in formulae booklets.

Laws of logarithms

$$\log_a x + \log_a y \equiv \log_a (xy)$$
$$\log_a x - \log_a y \equiv \log_a \left(\frac{x}{y}\right)$$
$$k \log_a x \equiv \log_a (x^k)$$

Trigonometry

In the triangle ABC

$$\frac{a}{\sin A} = \frac{b}{\sin B} = \frac{c}{\sin C}$$

$$\text{area} = \tfrac{1}{2} ab \sin C$$

Area

$$\text{area under a curve} = \int_a^b y \, dx \ (y \geqslant 0)$$

List of symbols and notation

The following notation will be used in all Edexcel mathematics examinations:

$\in$	is an element of
$\notin$	is not an element of
$\{x_1, x_2, ...\}$	the set with elements $x_1, x_2, ...$
$\{x: ...\}$	the set of all x such that ...
$n(A)$	the number of elements in set A
$\varnothing$	the empty set
ξ	the universal set
A'	the complement of the set A
$\mathbb{N}$	the set of natural numbers, $\{1, 2, 3, ...\}$
$\mathbb{Z}$	the set of integers, $\{0, \pm1, \pm2, \pm3, ...\}$
$\mathbb{Z}^+$	the set of positive integers, $\{1, 2, 3, ...\}$
$\mathbb{Z}_n$	the set of integers modulo n, $\{1, 2, 3, ..., n-1\}$
$\mathbb{Q}$	the set of rational numbers, $\left\{\dfrac{p}{q}: p \in \mathbb{Z}_u, q \in \mathbb{Z}^+\right\}$
$\mathbb{Q}^+$	the set of positive rational numbers, $\{x \in \mathbb{Q}: x > 0\}$
$\mathbb{Q}_0^+$	the set of positive rational numbers and zero, $\{x \in \mathbb{Q}: x \geqslant 0\}$
$\mathbb{R}$	the set of real numbers
$\mathbb{R}^+$	the set of positive real numbers, $\{x \in \mathbb{R}: x > 0\}$
$\mathbb{R}_0^+$	the set of positive real numbers and zero, $\{x \in \mathbb{R}: x \geqslant 0\}$
$\mathbb{C}$	the set complex numbers
(x, y)	the ordered pair x, y
$A \times B$	the cartesian products of sets A and B, ie $A \times B = \{(a, b): a \in A, b \in B\}$
$\subseteq$	is a subset of
$\subset$	is a proper subset of
$\cup$	union
$\cap$	intersection
$[a, b]$	the closed interval, $\{x \in \mathbb{R}: a \leqslant x \leqslant b\}$
$[a, b), [a, b[$	the interval, $\{x \in \mathbb{R}: a \leqslant x < b\}$
$(a, b],]a, b]$	the interval, $\{x \in \mathbb{R}: a < x \leqslant b\}$
$(a, b),]a, b[$	the open interval, $\{x \in \mathbb{R}: a < x < b\}$
$y \, R \, x$	y is related to x by the relation R
$y \sim x$	y is equivalent to x, in the context of some equivalence relation
$=$	is equal to
$\neq$	is not equal to
$\equiv$	is identical to or is congruent to
$\approx$	is approximately equal to
$\cong$	is isomorphic to
$\propto$	is proportional to
$<$	is less than
$\leqslant, \not>$	is less than or equal to, is not greater than

$>$	is greater than
$\geqslant$, $\nless$	is greater than or equal to, is not less than
∞	infinity
$p \wedge q$	p and q
$p \vee q$	p or q (or both)
$\sim p$	not p
$p \Rightarrow q$	p implies q (if p then q)
$p \Leftarrow q$	p is implied by q (if q then p)
$p \Leftrightarrow q$	p implies and is implied by q (p is equivalent to q)
$\exists$	there exists
$\forall$	for all
$a + b$	a plus b
$a - b$	a minus b
$a \times b$, ab, $a.b$	a multiplied by b
$a \div b$, $\dfrac{a}{b}$, a/b	a divided by b
$\displaystyle\sum_{i=1}^{n}$	$a_1 + a_2 + \ldots + a_n$
$\displaystyle\prod_{i=1}^{n}$	$a_1 \times a_2 \times \ldots \times a_n$
$\sqrt{a}$	the positive square root of a
$|a|$	the modulus of a
$n!$	n factorial
$\dbinom{n}{r}$	the binomial coefficient $\dfrac{n!}{r!(n-r)!}$ for $n \in \mathbb{Z}^+$
	$\dfrac{n(n-1)\ldots(n-r+1)}{r!}$ for $n \in \mathbb{Q}$
f(x)	the value of the function f at x
f$: A \rightarrow B$	f is a function under which each element of set A has an image in set B
f$: x \rightarrow y$	the function f maps the element x to the element y
f^{-1}	the inverse function of the function f
g $\circ$ f, gf	the composite function of f and g which is defined by (g $\circ$ f)(x) or gf(x) = g(f(x))
$\displaystyle\lim_{x \to a}$ f(x)	the limit of(x) of as x tends to a
Δx, δx	an increment of x
$\dfrac{dy}{dx}$	the derivative of y with respect to x
$\dfrac{d^n y}{dx^n}$	the nth derivative of y with respect to x
f$'(x)$, f$''(x)$, $\ldots$, f$^{(n)}(x)$	the first, second, $\ldots$, nth derivatives of f(x) with respect to x
$\displaystyle\int y \, dx$	the indefinite integral of y with respect to x
$\displaystyle\int_{b}^{a} y \, dx$	the definite integral of y with respect to x between the limits
$\dfrac{\partial V}{\partial x}$	the partial derivative of V with respect to x
$\dot{x}$, $\ddot{x}$, $\ldots$	the first, second, $\ldots$ derivatives of x with respect to t

e	base of natural logarithms
e^x, exp x	exponential function of x
$\log_a x$	logarithm to the base a of x
$\ln x$, $\log_e x$	natural logarithm of x
$\lg x$, $\log_{10} x$	logarithm of x to base 10
sin, cos, tan, cosec, sec, cot	the circular functions
arcsin, arccos, arctan, arccosec, arcsec, arccot	the inverse circular functions
sinh, cosh, tanh, cosech, sech, coth	the hyperbolic functions
arsinh, arcosh, artanh, arcosech, arsech, arcoth	the inverse hyperbolic functions
i, j	square root of -1
z	a complex number, $z = x + \mathrm{i}y$
Re z	the real part of z, Re $z = x$
Im z	the imaginary part of z, Im $z = y$
$\lvert z \rvert$	the modulus of z, $\lvert z \rvert = \sqrt{(x^2 + y^2)}$
arg z	the argument of z, arg $z = \theta$, $-\pi < \theta \leqslant \pi$
$z^\ast$	the complex conjugate of z, $x - \mathrm{i}y$
M	a matrix **M**
$\mathbf{M}^{-1}$	the inverse of the matrix **M**
$\mathbf{M}^{\mathrm{T}}$	the transpose of the matrix **M**
det **M** or $\lvert \mathbf{M} \rvert$	the determinant of the square matrix **M**
a	the vector **a**
$\overrightarrow{AB}$	the vector represented in magnitude and direction by the directed line segment AB
â	a unit vector in the direction of **a**
i, j, k	unit vectors in the direction of the cartesian coordinate axes
$\lvert \mathbf{a} \rvert$, a	the magnitude of **a**
$\lvert \overrightarrow{AB} \rvert$	the magnitude of $\overrightarrow{AB}$
a . b	the scalar product of **a** and **b**
$\mathbf{a} \times \mathbf{b}$	the vector product of **a** and **b**

Answers

Exercise 1A

1 a $4x^3 + 5x - 7$ **b** $7x^7 - 5x^4 + 9x^2 + x$

c $-2x^2 + 1$ **d** $-x^3 + 4x + \dfrac{6}{x}$

e $7x^4 - x^2 - \dfrac{4}{x}$ **f** $4x^3 - 2x^2 + 3$

g $3x - 4x^2 - 1$ **h** $2x^4 - \dfrac{x^2}{2}$

i $\dfrac{7x^2}{5} - \dfrac{x^3}{5} - \dfrac{2}{5x}$ **j** $2x - 3x^3 + 1$

k $\dfrac{x^7}{2} - \dfrac{9x^3}{2} - \dfrac{3}{x}$ **l** $3x^8 + 2x^3 + \dfrac{2}{3x}$

2 a $x + 3$ **b** $x + 4$ **c** $x + 3$

d $x + 7$ **e** $x + 5$ **f** $x + 4$

g $\dfrac{x - 4}{x - 3}$ **h** $\dfrac{x + 2}{x + 4}$ **i** $\dfrac{x + 4}{x - 6}$

j $\dfrac{2x + 3}{x - 5}$ **k** $\dfrac{2x - 3}{x + 1}$ **l** $\dfrac{x - 2}{x + 2}$

m $\dfrac{2x + 1}{x - 2}$ **n** $\dfrac{x + 4}{3x + 1}$ **o** $\dfrac{2x + 1}{2x - 3}$

Exercise 1B

1 a $x^2 + 5x + 3$ **b** $x^2 + 6x + 1$ **c** $x^2 + x - 9$

d $x^2 + 4x - 2$ **e** $x^2 - 3x + 7$ **f** $x^2 + 4x + 5$

g $x^2 - 3x + 2$ **h** $x^2 - 2x + 6$ **i** $x^2 - 3x - 2$

j $x^2 + 2x + 8$

2 a $6x^2 + 3x + 2$ **b** $4x^2 + x - 5$

c $3x^2 + 2x - 2$ **d** $3x^2 + 4x + 8$

e $2x^2 - 2x - 3$ **f** $2x^2 - 3x - 4$

g $-3x^2 + 5x - 7$ **h** $-2x^2 - 3x + 5$

i $-5x^2 + 3x + 5$ **j** $-4x^2 + x - 1$

3 a $x^3 + 3x^2 - 4x + 1$ **b** $x^3 + 6x^2 - 5x - 4$

c $4x^3 + 2x^2 - 3x - 5$ **d** $3x^3 + 5x^2 - 3x + 2$

e $-3x^3 + 3x^2 - 4x - 7$

f $3x^4 + 2x^3 - 8x^2 + 2x - 5$

g $6x^4 - x^3 - 2x^2 - 5x - 2$

h $-5x^4 + 2x^3 + 4x^2 - 3x + 7$

i $2x^5 - 3x^4 + 2x^3 - 8x^2 + 4x + 6$

j $-x^5 + x^4 - x^3 + x^2 - 2x + 1$

Exercise 1C

1 a $x^2 - 2x + 5$ **b** $2x^2 - 6x + 1$

c $-3x^2 + 12x + 2$

2 a $x^2 + 4x + 12$ **b** $2x^2 - x + 5$

c $-3x^2 + 5x + 10$

3 a $x^2 - 5$ **b** $2x^2 + 7$ **c** $-3x^2 - 4$

4 a -8 **b** -7 **c** -12

7 $(x + 4)(5x^2 - 20x + 7)$

8 $3x^2 + 6x + 4$

9 $x^2 + x + 1$

10 $x^3 - 2x^2 + 4x - 8$

Exercise 1D

2 $(x - 1)(x + 3)(x + 4)$

3 $(x + 1)(x + 7)(x - 5)$

4 $(x - 5)(x - 4)(x + 2)$

5 $(x - 2)(2x - 1)(x + 4)$

6 a $(x + 1)(x - 5)(x - 6)$

b $(x - 2)(x + 1)(x + 2)$

c $(x - 5)(x + 3)(x - 2)$

7 a $(x - 1)(x + 3)(2x + 1)$ **b** $(x - 3)(x - 5)(2x - 1)$

c $(x + 1)(x + 2)(3x - 1)$ **d** $(x + 2)(2x - 1)(3x + 1)$

e $(x - 2)(2x - 5)(2x + 3)$

8 2

9 -16

10 $p = 3, q = 7$

Exercise 1E

1 a 27 **b** -6 **c** 0 **d** 1

e $2\frac{1}{4}$ **f** 8 **g** 14 **h** 0

i $-15\frac{1}{3}$ **j** -20.52

2 -1

3 18

4 30

7 -9

8 $8\frac{8}{27}$

9 $a = 5, b = -8$

10 $p = 8, q = 3$

Mixed exercise 1F

1 a $x^3 - 7$ **b** $\dfrac{x + 4}{x - 1}$ **c** $\dfrac{2x - 1}{2x + 1}$

2 $3x^2 + 5$

3 $2x^2 - 2x + 5$

4 $A = 2, B = 4, C = -5$

5 $p = 1, q = 3$

6 $(x - 2)(x + 4)(2x - 1)$

7 7

8 $7\frac{1}{4}$

9 a $p = -1, q = -15$ **b** $(x + 3)(2x - 5)$

10 a $r = 3, s = 0$ **b** $1\frac{13}{27}$

11 a $(x - 1)(x + 5)(2x + 1)$ **b** $-5, -\frac{1}{2}, 1$

12 -2

13 -18

14 $2, -\dfrac{3}{2} \pm \dfrac{\sqrt{5}}{2}$

15 $\frac{1}{2}, 3$

Exercise 2A

1 a $15.2\,\text{cm}$ **b** $9.57\,\text{cm}$ **c** $8.97\,\text{cm}$ **d** $4.61\,\text{cm}$

2 a $x = 84, y = 6.32$

b $x = 13.5, y = 16.6$

c $x = 85, y = 13.9$

d $x = 80, y = 6.22$ (Isosceles $\triangle$)

e $x = 6.27, y = 7.16$

f $x = 4.49, y = 7.49$ (right-angled)

3 a $1.41\,\text{cm}$ ($\sqrt{2}\,\text{cm}$)   **b** $1.93\,\text{cm}$

4 a $6.52\,\text{km}$ **b** $3.80\,\text{km}$

5 a $7.31\,\text{cm}$ **b** $1.97\,\text{cm}$

Exercise 2B

1 a 36.4 **b** 35.8 **c** 40.5 **d** 130

2 a 48.1 **b** 45.6 **c** 14.8

d 48.7 **e** 86.5 **f** 77.4

3 $\angle QPR = 50.6°$, $\angle PQR = 54.4°$

4 a $x = 43.2$, $y = 5.02$ **b** $x = 101$, $y = 15.0$
   c $x = 6.58$, $y = 32.1$ **d** $x = 54.6$, $y = 10.3$
   e $x = 21.8$, $y = 3.01$ **f** $x = 45.9$, $y = 3.87$

5 Using the sine rule, $x = \dfrac{4\sqrt{2}}{2 + \sqrt{2}}$; rationalising

$x = \dfrac{4\sqrt{2}(2 - \sqrt{2})}{2} = 4\sqrt{2} - 4 = 4(\sqrt{2} - 1).$

Exercise 2C

1 a $70.5°$, $109°$ $(109.5°)$

2 a $x = 74.6$, $y = 65.4$
   $x = 105$, $y = 34.6$
   b $x = 59.8$, $y = 48.4$
   $x = 120$, $y = 27.3$
   c $x = 56.8$, $y = 4.37$
   $x = 23.2$, $y = 2.06$

3 a 5 cm $(\angle ACB = 90°)$ **b** $24.6°$
   c $45.6°$, $134(.4)°$

4 2.96 cm

5 In one triangle $\angle ABC = 101°$ $(100.9°)$; in the other $\angle BAC = 131°$ $(130.9°)$.

Exercise 2D

1 a 3.19 cm   **b** 1.73 cm $(\sqrt{3}$ cm$)$
   c 9.85 cm   **d** 4.31 cm
   e 6.84 cm (isosceles)   **f** 9.80 cm

2 11.2 km

3 302 yards $(301.5...)$

4 4.4

5 42

6 b Minimum $AC^2 = 60.75$; it occurs for $x = \tfrac{1}{2}$.

Exercise 2E

1 a $108(.2)°$ **b** $90°$ **c** $60°$
   d $52.6°$ **e** $137°$ **f** $72.2°$

2 $128.5°$ or $31.5°$ $(\angle BAC = 48.5°)$

3 $\angle ACB = 22.3°$

4 $\angle ABC = 108(.4)°$

5 $104°$ $(104.48°)$

6 b 3.5

Exercise 2F

1 a $x = 37.7$, $y = 86.3$, $z = 6.86$
   b $x = 48$, $y = 19.5$, $z = 14.6$
   c $x = 30$, $y = 11.5$, $z = 11.5$
   d $x = 21.0$, $y = 29.0$, $z = 8.09$
   e $x = 93.8$, $y = 56.3$, $z = 29.9$
   f $x = 97.2$, $y = 41.4$, $z = 41.4$
   g $x = 45.3$, $y = 94.7$, $z = 14.7$
   or $x = 134.7$, $y = 5.27$, $z = 1.36$
   h $x = 7.07$, $y = 73.7$, $z = 61.2$
   or $x = 7.07$, $y = 106$, $z = 28.7$
   i $x = 49.8$, $y = 9.39$, $z = 37.0$

2 a $\angle ABC = 108°$, $\angle ACB = 32.4°$, $AC = 15.1$ cm
   b $\angle BAC = 41.5°$, $\angle ABC = 28.5°$, $AB = 9.65$ cm

3 a 8 km   **b** $060°$

4 107 km

5 12 km

6 a 5.44 **b** 7.95 **c** 36.8

7 a $AB + BC > AC \Rightarrow x + 6 > 7 \Rightarrow x > 1$;
   $AC + AB > BC \Rightarrow 11 > x + 2 \Rightarrow x < 9$
   b i $x = 6.08$ from $x^2 = 37$,
    ii $x = 7.23$ from
     $x^2 - 4(\sqrt{2} - 1)x - (29 + 8\sqrt{2}) = 0$

8 $x = 4$

9 $AC = 1.93$ cm

10 b $\tfrac{1}{2}$

11 $4\sqrt{10}$

12 $AC = 1\tfrac{2}{3}$ cm and $BC = 6\tfrac{1}{3}$ cm

Exercise 2G

1 a 23.7 cm²   **b** 4.31 cm²   **c** 20.2 cm²

2 a $x = 41.8$ or $138(.2)$
   b $x = 26.7$ or $153(.3)$
   c $x = 60$ or 120

3 $275(.3)$ m (third side $= 135.3$ m)

4 3.58

5 b Maximum $A = 3\tfrac{1}{16}$, when $x = 1\tfrac{1}{2}$

6 b 2.11

Mixed exercise 2H

1 a $155°$
   b 13.7 cm

2 a $x = 49.5$, area $= 1.37$ cm²
   b $x = 55.2$, area $= 10.6$ cm²
   c $x = 117$, area $= 6.66$ cm²

3 6.50 cm²

4 a 36.1 cm²   **b** 12.0 cm²

5 a 5 **b** $\dfrac{25\sqrt{3}}{2}$ cm²

7 b $1\tfrac{1}{2}$ cm²

8 a 4 **b** $\dfrac{15\sqrt{3}}{4}$ (6.50) cm²

Exercise 3A

1 a

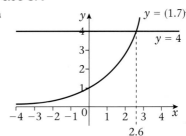

   b $x \approx 2.6$

2 a

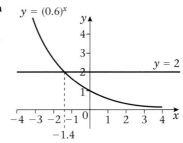

   b $x \approx -1.4$

3

Exercise 3B

1 a $\log_4 256 = 4$ **b** $\log_3 (\frac{1}{9}) = -2$
  c $\log_{10} 1\,000\,000 = 6$ **d** $\log_{11} 11 = 1$
  e $\log_{0.2} 0.008 = 3$

2 a $2^4 = 16$ **b** $5^2 = 25$
  c $9^{\frac{1}{2}} = 3$ **d** $5^{-1} = 0.2$
  e $10^5 = 100\,000$

3 a 3   **b** 2   **c** 7   **d** 1   **e** 6
  f $\frac{1}{2}$ **g** -1 **h** -2 **i** 10   **j** -2

4 a 625   **b** 9   **c** 7   **d** 2

Exercise 3C

1 1.30   **2** 0.602   **3** 3.85   **4** -0.105
5 1.04   **6** 1.55   **7** -0.523 **8** 3.00

Exercise 3D

1 a $\log_2 21$ **b** $\log_2 9$ **c** $\log_5 80$
  d $\log_6 (\frac{64}{81})$ **e** $\log_{10} 120$

2 a $\log_2 8 = 3$ **b** $\log_6 36 = 2$ **c** $\log_{12} 144 = 2$
  d $\log_8 2 = \frac{1}{3}$ **e** $\log_{10} 10 = 1$

3 a $3\log_a x + 4\log_a y + \log_a z$
  b $5\log_a x - 2\log_a y$
  c $2 + 2\log_a x$
  d $\log_a x + \frac{1}{2}\log_a y - \log_a z$
  e $\frac{1}{2} + \frac{1}{2}\log_a x$

Exercise 3E

1 a 6.23   **b** 2.10   **c** 0.431   **d** 1.66
  e -3.22 **f** 1.31   **g** -3.24 **h** -0.0617
  i 1.42   **j** -0.542

2 a 0, 2.32   **b** 1.26, 2.18   **c** 1.21
  d 0.631   **e** 0.565, 0.712

Exercise 3F

1 a 2.460   **b** 3.465   **c** 4.248
  d 0.458   **e** 0.774
2 a 1.27   **b** 2.09   **c** 0.721
3 a $\frac{1}{2}$, 512   **b** $\frac{1}{16}, \frac{1}{4}$ **c** 2.52

Mixed exercise 3G

1 $x = -1, x = 0$
2 a $2\log_a p + \log_a q$ **b** $\log_a p = 4, \log_a q = 1$
3 a $\frac{1}{4}p$ **b** $\frac{3}{4}p + 1$
4 a 9   **b** 12   **c** $\frac{1}{9}$, 9
5 b 2.32
6 $x = \frac{3}{22}, y = \frac{24}{11}$
7 $\frac{1}{3}$, 9
8 $-\frac{1}{3}, -2$
9 (4, 16) or (16, 4)
11 b $x = \dfrac{\sqrt{3}}{4}, y = \dfrac{\sqrt{3}}{2}$
12 b $\alpha = \frac{1}{4}, \beta = \frac{3}{2}$ **d** 0.585

Exercise 4A

1 a $(5, 5)$ **b** $(6, 4)$ **c** $(-1, 4)$
  d $(0, 0)$ **e** $(1, 4)$ **f** $(2, 1)$
  g $(-8, \frac{3}{2})$ **h** $(4a, 0)$ **i** $(3p, 2q)$
  j $\left(\dfrac{3s}{2}, -3t\right)$ **k** $\left(-\dfrac{u}{2}, -v\right)$ **l** $(2a, a - b)$
  m $(3\sqrt{2}, 4)$ **n** $\left(2\sqrt{3}, \dfrac{5\sqrt{5}}{2}\right)$ **o** $(2\sqrt{2}, \sqrt{2} + 3\sqrt{3})$

2 $(\frac{3}{2}, 7)$
3 $\left(\dfrac{3a}{5}, \dfrac{b}{4}\right)$
4 $(\frac{3}{2}, 3)$
5 $(\frac{1}{8}, \frac{5}{3})$
6 $(3, -\frac{7}{2})$
7 $(10, 5)$
8 $(-7a, 17a)$
9 $p = 8, q = 7$
10 $a = -2, b = 4$

Exercise 4B

1 $y = -x + 7$
2 $2x - y - 8 = 0$
3 a $(10, -10)$
  b $y = \frac{3}{4}x - \frac{35}{2}$
4 $y = -\frac{7}{3}x - 50$
6 $8x + 6y - 5 = 0$
7 $(-3, 2)$
8 $(8, 0)$
9 a i $y = 2x$ **ii** $y = -x + 9$
  b $(3, 6)$
10 $(-3, 6)$

Exercise 4C

1 a 10   **b** 13   **c** 5
  d $\sqrt{5}$ **e** $2\sqrt{10}$ **f** $\sqrt{106}$
  g $\sqrt{113}$ **h** $a\sqrt{53}$ **i** $3b\sqrt{5}$
  j $5c$ **k** $d\sqrt{61}$ **l** $2e\sqrt{5}$
  m $\sqrt{10}$ **n** $5\sqrt{3}$ **o** $4\sqrt{2}$

2 10
4 $\sqrt{10}$
5 $(3, 6)$
7 a i $2\sqrt{5}$ **ii** 5
  b $\sqrt{10}$
8 $(2, 3)$
9 15
10 b 50   **c** $(3, 6)$

Exercise 4D

1 a $(x - 3)^2 + (y - 2)^2 = 16$
  b $(x + 4)^2 + (y - 5)^2 = 36$
  c $(x - 5)^2 + (y + 6)^2 = 12$
  d $(x - 2a)^2 + (y - 7a)^2 = 25a^2$
  e $(x + 2\sqrt{2})^2 + (y + 3\sqrt{2})^2 = 1$

2 a $(-5, 4), 9$ **b** $(7, 1), 4$
  c $(-4, 0), 5$ **d** $(-4a, -a), 12a$
  e $(3\sqrt{5}, -\sqrt{5}), 3\sqrt{3}$

4 $(x - 8)^2 + (y - 1)^2 = 25$
5 $(x - \frac{3}{2})^2 + (y - 4)^2 = \frac{65}{4}$
6 $\sqrt{5}$
8 a $3\sqrt{10}$
9 a $(x - 4)^2 + (y - 6)^2 = 73$ **b** $3x + 8y + 13 = 0$
10 a $(0, -17), (17, 0)$ **b** 144.5

Exercise 4E

1 $(7, 0), (-5, 0)$
2 $(0, 2), (0, -8)$
3 $a = -2, 8$ $b = -8, 2$
4 $(6, 10), (-2, 2)$
5 $(4, -9), (-7, 2)$
9 $(0, -2), (4, 6)$
10 a 13   **b** 1, 5

Mixed exercise 4F

1 $(x - 2)^2 + (y + 4)^2 = 20$
2 a $2\sqrt{29}$ **b** 12
3 $(-1, 0), (11, 0)$
4 $m = 7 - \sqrt{105}, n = 7 + \sqrt{105}$
5 4
6 $x + y + 10 = 0$
7 a $a = 3, b = \sqrt{91} - 8$
   b $y = \left(\dfrac{8 - \sqrt{91}}{3}\right)(x - 3)$
8 a $c = 10, d = \sqrt{165} + 5$ **b** $5(\sqrt{165} + 5)$
9 a $p = 0, q = 24$ **b** $(0, 49), (0, -1)$
11 $(1, 3)$
12 $(x - 2)^2 + (y + 2)^2 = 61$
13 a i $y = -4x - 4$ **ii** $x = -2$ **b** $(-2, 4)$
14 b $\sqrt{106}$
15 a $(4, 0), (0, 12)$ **b** $(2, 6)$
   c $(x - 2)^2 + (y - 6)^2 = 40$
16 $(x + 2)^2 + y^2 = 34$
17 60
19 a $r = 2$
20 a $p = 1, q = 4$ **b i** $-\frac{2}{3}$ **ii** $\frac{3}{2}$

Exercise 5A

1 a $x^4 + 4x^3y + 6x^2y^2 + 4xy^3 + y^4$
   b $p^5 + 5p^4q + 10p^3q^2 + 10p^2q^3 + 5pq^4 + q^5$
   c $a^3 - 3a^2b + 3ab^2 - b^3$
   d $x^3 + 12x^2 + 48x + 64$
   e $16x^4 - 96x^3 + 216x^2 - 216x + 81$
   f $a^5 + 10a^4 + 40a^3 + 80a^2 + 80a + 32$
   g $81x^4 - 432x^3 + 864x^2 - 768x + 256$
   h $16x^4 - 96x^3y + 216x^2y^2 - 216xy^3 + 81y^4$
2 a 16 **b** -10 **c** 8 **d** 1280
   e 160 **f** -2 **g** 40 **h** -96
3 d $1 + 9x + 30x^2 + 44x^3 + 24x^4$
4 $8 + 12y + 6y^2 + y^3, 8 + 12x - 6x^2 - 11x^3 + 3x^4 + 3x^5 - x^6$
5 -143
6 ± 3
7 $\frac{5}{2}, -1$
8 $\frac{3}{4}$

Exercise 5B

1 a 24 **b** 720 **c** 56
   d 10 **e** 6 **f** 28
   g 10 **h** 20 **i** 10
   j 15 **k** 56 **l** $\dfrac{n(n - 1)(n - 2)}{6}$
2 a 1 **b** 4 **c** 6 **d** 4 **e** 1
3 a $\binom{3}{0} \binom{3}{1} \binom{3}{2} \binom{3}{3}$ **b** $\binom{5}{0} \binom{5}{1} \binom{5}{2} \binom{5}{3} \binom{5}{4} \binom{5}{5}$
4 a Selecting a group of 4 from 6 creates a group of 2.
   b $^6C_2 = 15, \binom{6}{4} = 15$

Exercise 5C

1 a $16x^4 + 32x^3y + 24x^2y^2 + 8xy^3 + y^4$
   b $p^5 - 5p^4q + 10p^3q^2 - 10p^2q^3 + 5pq^4 - q^5$
   c $1 + 8x + 24x^2 + 32x^3 + 16x^4$
   d $81 + 108x + 54x^2 + 12x^3 + x^4$
   e $1 - 2x + \frac{3}{2}x^2 - \frac{1}{2}x^3 + \frac{1}{16}x^4$
   f $256 - 256x + 96x^2 - 16x^3 + x^4$
   g $32x^5 - 240x^4y + 720x^3y^2 - 1080x^2y^3$
       $+ 810xy^4 - 243y^5$
   h $x^6 + 12x^5 + 60x^4 + 160x^3 + 240x^2 + 192x + 64$

2 a $90x^3$ **b** $80x^3y^2$ **c** $-20x^3$
   d $720x^3$ **e** $120x^3$ **f** $-4320x^3$
   g $1140x^3$ **h** $-241\,920x^3$
3 a $1 + 10x + 45x^2 + 120x^3$
   b $1 - 10x + 40x^2 - 80x^3$
   c $1 + 18x + 135x^2 + 540x^3$
   d $256 - 1024x + 1792x^2 - 1792x^3$
   e $1024 - 2560x + 2880x^2 - 1920x^3$
   f $2187 - 5103x + 5103x^2 - 2835x^3$
   g $x^8 + 16x^7y + 112x^6y^2 + 448x^5y^3$
   h $512x^9 - 6912x^8y + 41\,472x^7y^2 - 145\,152x^6y^3$
4 $a = \pm\frac{1}{2}$
5 $b = -2$
6 $1, 5 \pm \dfrac{\sqrt{105}}{8}$
7 $1 - 0.6x + 0.15x^2 - 0.02x^3, 0.941\,48,$ accurate to 5 dp
8 $1024 + 1024x + 460.8x^2 + 122.88x^3, 1666.56,$
   accurate to 3 sf

Exercise 5D

1 a $1 + 8x + 28x^2 + 56x^3$
   b $1 - 12x + 60x^2 - 160x^3$
   c $1 + 5x + \frac{45}{4}x^2 + 15x^3$
   d $1 - 15x + 90x^2 - 270x^3$
   e $128 + 448x + 672x^2 + 560x^3$
   f $27 - 54x + 36x^2 - 8x^3$
   g $64 - 576x + 2160x^2 - 4320x^3$
   h $256 + 256x + 96x^2 + 16x^3$
   i $128 + 2240x + 16\,800x^2 + 70\,000x^3$
3 $a = 162, b = 135, c = 0$
4 a $p = 5$ **b** -10 **c** -80
5 $1 + 16x + 112x^2 + 448x^3, 1.171\,648,$ accurate to 4 sf

Mixed exercise 5E

1 a $p = 16$ **b** 270 **c** -1890
2 a $A = 8192, B = -53\,248, C = 159\,744$
3 a $1 - 20x + 180x^2 - 960x^3$
   b $0.817\,04, x = 0.01$
4 a $1024 - 153\,60x + 103\,680x^2 - 414\,720x^3$
   b 880.35
5 a $81 + 216x + 216x^2 + 96x^3 + 16x^4$
   b $81 - 216x + 216x^2 - 96x^3 + 16x^4$
   c 1154
6 a $n = 8$ **b** $\frac{35}{8}$
7 a $81 + 1080x + 5400x^2 + 12\,000x^3 + 10\,000x^4$
   b $1\,012\,054\,108\,081, x = 100$
8 a $1 + 24x + 264x^2 + 1760x^3$
   b $1.268\,16$
   c $1.268\,241\,795$
   d $0.006\,45\%$ (3 sf)
9 $x^5 - 5x^3 + 10x - \dfrac{10}{x} + \dfrac{5}{x^3} - \dfrac{1}{x^5}$
10 b $\dfrac{4096}{729} + \dfrac{2048}{81}x + \dfrac{1280}{27}x^2 + \dfrac{1280}{27}x^3$
11 a $64 + 192x + 240x^2 + 160x^3 + 60x^4 + 12x^5 + x^6$
   b $k = 1560$
12 a $k = 1.25$ **b** 3500
13 a $A = 64, B = 160, C = 20$ **b** $x = \pm\sqrt{\frac{3}{2}}$
14 a $p = 1.5$ **b** 50.625

Exercise 6A

1 a $9°$ **b** $12°$ **c** $75°$
   d $90°$ **e** $140°$ **f** $210°$
   g $225°$ **h** $270°$ **i** $540°$

2 a 26.4° **b** 57.3° **c** 65.0° **d** 99.2°
e 143.2° **f** 179.9° **g** 200.0°
3 a 0.479 **b** 0.156 **c** 1.74
d 0.909 **e** −0.897
4 a $\dfrac{2\pi}{45}$ **b** $\dfrac{\pi}{18}$ **c** $\dfrac{\pi}{8}$ **d** $\dfrac{\pi}{6}$
e $\dfrac{\pi}{4}$ **f** $\dfrac{\pi}{3}$ **g** $\dfrac{5\pi}{12}$ **h** $\dfrac{4\pi}{9}$
i $\dfrac{5\pi}{8}$ **j** $\dfrac{2\pi}{3}$ **k** $\dfrac{3\pi}{4}$ **l** $\dfrac{10\pi}{9}$
m $\dfrac{4\pi}{3}$ **n** $\dfrac{3\pi}{2}$ **o** $\dfrac{7\pi}{4}$ **p** $\dfrac{11\pi}{6}$
5 a 0.873 **b** 1.31 **c** 1.75
d 2.79 **e** 4.01 **f** 5.59

Exercise 6B

1 a i 2.7 **ii** 2.025 **iii** 7.5π (23.6)
b i $16\frac{2}{3}$ **ii** 1.8 **iii** 3.6
c i $1\frac{1}{3}$ **ii** 0.8 **iii** 2
2 $\dfrac{10\pi}{3}$ cm
3 2π
4 $5\sqrt{2}$ cm
5 a 10.4 cm **b** $1\frac{1}{4}$
6 7.5
7 0.8
8 a $\dfrac{\pi}{3}$ **b** $\left(6 + \dfrac{4\pi}{3}\right)$ cm
9 6.8 cm
10 a $(R - r)$ cm **b** 2.43

Exercise 6C

1 a 19.2 cm² **b** 6.75π cm² **c** 1.296π cm²
d 38.3 cm² **e** $5\frac{1}{3}\pi$ cm² **f** 5 cm²
2 a 4.47 **b** 3.96 **c** 1.98
3 12 cm²
4 b 120 cm²
5 $40\frac{2}{3}$ cm
6 a 12 **c** 1.48 cm²
8 38.7 cm²
9 8.88 cm²
10 a 1.75 cm² **b** 25.9 cm² **c** 25.9 cm²
11 9 cm²
12 b 28 cm
13 78.4 ($\theta = 0.8$)
14 b 34.1 m²

Mixed exercise 6D

1 a $\dfrac{\pi}{3}$ **b** 8.56 cm²
2 a 120 cm² **b** 2.16 **c** 161.07 cm²
3 a 1.839 **b** 11.03
4 a $\dfrac{p}{r}$ **c** 12.206 cm²
d $1.106 < \theta < 1.150$
5 a 1.28 **b** 16 **c** 1 : 3.91
6 b $(3\pi + 12)$ cm
c $\left(18 + \dfrac{3\pi}{2}\right)$ cm
d 12.9 mm
7 c f(2.3) = −0.033, f(2.32) = +0.034
8 b i 6.77 **ii** 15.7 **iii** 22.5
9 b 15.7 **c** 5.025 cm²
11 a $2\sqrt{3}$ cm **b** 2π cm²

12 b i 80.9 m **ii** 26.7 m **iii** 847 m²
13 c 20.7 cm²
14 b 16π cm **c** 177 cm²

Exercise 7A

1 a Geometric $r = 2$ **b** Not geometric
c Not geometric **d** Geometric $r = 3$
e Geometric $r = \frac{1}{2}$ **f** Geometric $r = -1$
g Geometric $r = 1$ **h** Geometric $r = -\frac{1}{4}$
2 a 135, 405, 1215 **b** −32, 64, −128
c 7.5, 3.75, 1.875 **d** $\frac{1}{64}, \frac{1}{256}, \frac{1}{1024}$
e p^3, p^4, p^5 **f** $-8x^4, 16x^5, -32x^6$
3 a $3\sqrt{3}$ **b** $9\sqrt{3}$

Exercise 7B

1 a 486, 39 366, $2 \times 3^{n-1}$
b $\dfrac{25}{8}, \dfrac{25}{128}, \dfrac{100}{2^{n-1}}$
c −32, −512, $(-2)^{n-1}$
d 1.610 51, 2.357 95, $(1.1)^{n-1}$
2 10, 6250
3 $a = 1$, $r = 2$
4 $\pm\frac{1}{8}$
5 −6 (from $x = 0$), 4 (from $x = 10$)

Exercise 7C

1 a 220 **b** 242 **c** 266 **d** 519
2 57.7, 83.2
3 £18 000, after 7.88 years
4 34
5 11th term
6 59 days
7 20.15 years
8 11.2 years

Exercise 7D

1 a 255 **b** 63.938 (3 dp)
c −728 **d** $546\frac{2}{3}$
e 5460 **f** 19 680
g 5.994 (3 dp) **h** 44.938 (3 dp)
2 $\frac{5}{4}, -\frac{9}{4}$
3 $2^{64} - 1 = 1.84 \times 10^{19}$
4 a £49 945.41 **b** £123 876.81
5 a 2.401 **b** 48.8234
6 19 terms
7 22 terms
8 26 days, 98.5 miles on the 25th day
9 25 years

Exercise 7E

1 a $\dfrac{10}{9}$ **b** Doesn't exist
c $6\frac{2}{3}$ **d** Doesn't exist
e Doesn't exist **f** $4\frac{1}{2}$
g Doesn't exist **h** 90
i $\dfrac{1}{1 - r}$ if $|r| < |$ **j** $\dfrac{1}{1 + 2x}$ if $|x| < \frac{1}{2}$
2 $\frac{2}{3}$
3 $-\frac{2}{3}$
4 20
5 $\frac{40}{3} = 13\frac{1}{3}$
6 $\frac{23}{99}$
7 4
8 40 m

9 $r < 0$ because $S_\infty < S_3$, $a = 12$, $r = -\frac{1}{2}$

10 $r = \pm\dfrac{\sqrt{2}}{3}$

Mixed exercise 7F

1 a Not geometric   **b** Geometric $r = 1.5$
   c Geometric $r = \frac{1}{2}$ **d** Geometric $r = -2$
   e Not geometric   **f** Geometric $r = 1$

2 a 0.8235 (4 dp), $10 \times (0.7)^{n-1}$
   b 640, $5 \times 2^{n-1}$
   c -4, $4 \times (-1)^{n-1}$
   d $-\frac{3}{128}$, $3 \times (-\frac{1}{2})^{n-1}$

3 a 4092 **b** 19.98 (2 dp)
   c 50 **d** 3.33 (2 dp)

4 a 9 **b** $\frac{8}{3}$
   c Doesn't converge   **d** $\frac{16}{3}$

5 b 60.75 **c** 182.25 **d** 3.16

6 b 200 **c** $333\frac{1}{3}$ **d** 8.95×10^{-4}

7 a 76, 60.8 **b** 0.876 **c** 367 **d** 380

8 a 1, $\frac{1}{3}$, $-\frac{1}{9}$

9 a 0.8 **b** 10 **c** 50 **d** 0.189 (3 sf)

10 a $V = 2000 \times 0.85^a$ **b** £8874.11
   c 9.9 years

11 a $-\frac{1}{2}$ **b** $\frac{3}{4}$, -2 **c** 14 **d** 867.62

12 b £12 079.98

13 a 1 hour 21 mins   **c** 3 hours 25 mins

14 a 136 litres   **c** 872

15 b 2015   **c** £23 700 (£23 657)

16 25 years

Exercise 8A

1 a
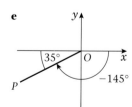

   b
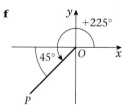

   c
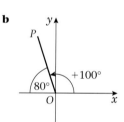

   d
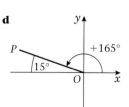

   e
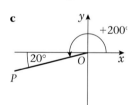

   f
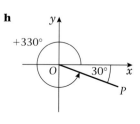

   g
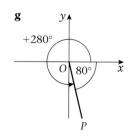

   h
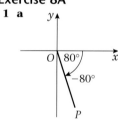

   i
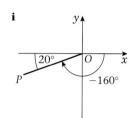

   j
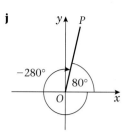

   k
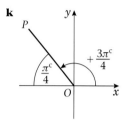

   l
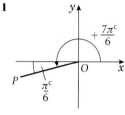

   m
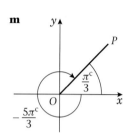

   n
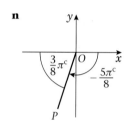

   o
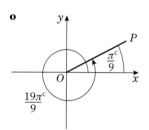

2 a First   **b** Second   **c** Second   **d** Third
   e Third   **f** Second   **g** First   **h** Fourth

Exercise 8B

1 a -1 **b** 1 **c** 0 **d** -1
   e -1 **f** 0 **g** 0 **h** 0
   i 0 **j** 0

2 a -1 **b** -1 **c** 0 **d** -1
   e 1 **f** -1 **g** 0 **h** 0
   i 0 **j** 0

Exercise 8C

1 a $-\sin 60°$ **b** $-\sin 80°$ **c** $\sin 20°$
   d $-\sin 60°$ **e** $\sin 80°$ **f** $-\cos 70°$
   g $-\cos 80°$ **h** $\cos 50°$ **i** $-\cos 20°$
   j $-\cos 5°$ **k** $-\tan 80°$ **l** $-\tan 35°$
   m $-\tan 30°$ **n** $\tan 5°$ **o** $\tan 60°$
   p $-\sin\dfrac{\pi}{6}$ **q** $-\cos\dfrac{\pi}{3}$ **r** $-\cos\dfrac{\pi}{4}$
   s $\tan\dfrac{2\pi}{5}$ **t** $-\tan\dfrac{\pi}{3}$ **u** $\sin\dfrac{\pi}{16}$
   v $\cos\dfrac{2\pi}{5}$ **w** $-\sin\dfrac{\pi}{7}$ **x** $-\tan\dfrac{\pi}{8}$

2 a $-\sin\theta$ **b** $-\sin\theta$ **c** $-\sin\theta$
   d $\sin\theta$ **e** $-\sin\theta$ **f** $\sin\theta$
   g $-\sin\theta$ **h** $-\sin\theta$ **i** $\sin\theta$

3 a $-\cos\theta$ **b** $-\cos\theta$ **c** $\cos\theta$
d $-\cos\theta$ **e** $\cos\theta$ **f** $-\cos\theta$
g $-\tan\theta$ **h** $-\tan\theta$ **i** $\tan\theta$
j $\tan\theta$ **k** $-\tan\theta$ **l** $\tan\theta$

4 $\sin\theta$ and $\tan\theta$ are odd functions
$\cos\theta$ is an even function

Exercise 8D

1 a $\dfrac{\sqrt{2}}{2}$ **b** $-\dfrac{\sqrt{3}}{2}$ **c** $-\dfrac{1}{2}$

d $\dfrac{\sqrt{3}}{2}$ **e** $\dfrac{\sqrt{3}}{2}$ **f** $-\dfrac{1}{2}$

g $\dfrac{1}{2}$ **h** $-\dfrac{\sqrt{2}}{2}$ **i** $-\dfrac{\sqrt{3}}{2}$

j $-\dfrac{\sqrt{2}}{2}$ **k** -1 **l** -1

m $\dfrac{\sqrt{3}}{3}$ **n** $-\sqrt{3}$ **o** $\sqrt{3}$

2

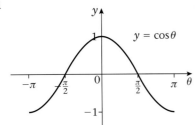

$\angle CAB = (90° - \theta)$

So $\cos(90° - \theta) = \dfrac{AC}{AB} = \dfrac{b}{c} = \sin\theta$

$\sin(90° - \theta) = \dfrac{BC}{AB} = \dfrac{a}{c} = \cos\theta$

$\tan(90° - \theta) = \dfrac{BC}{AC} = \dfrac{a}{b} = \dfrac{1}{\left(\dfrac{b}{a}\right)} = \dfrac{1}{\tan\theta}$

Exercise 8E

1

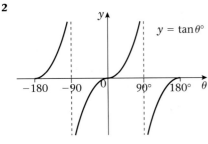

2

3

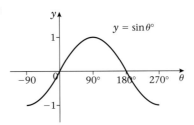

Exercise 8F

1 a i $1, x = 0$ **ii** $-1, x = 180$
b i $4, x = 90$ **ii** $-4, x = 270$
c i $1, x = 0$ **ii** $-1, x = 180$
d i $4, x = 90$ **ii** $2, x = 270$
e i $1, x = 270$ **ii** $-1, x = 90$
f i $1, x = 30$ **ii** $-1, x = 90$

2

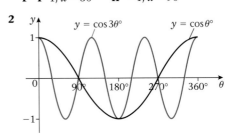

3 a The graph of $y = -\cos\theta$ is the graph of $y = \cos\theta$ reflected in the θ-axis.

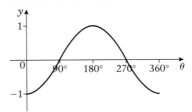

Meets θ-axis at $(90°, 0)$, $(270°, 0)$
Meets y-axis at $(0°, -1)$
Maximum at $(180°, 1)$
Minimum at $(0°, -1)$ and $(360°, -1)$

b The graph of $y = \frac{1}{3}\sin\theta$ is the graph of $y = \sin\theta$ stretched by a scale factor $\frac{1}{3}$ in the y direction.

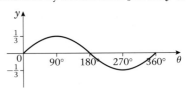

Meets θ-axis at $(0°, 0)$, $(180°, 0)$, $(360°, 0)$
Meets y-axis at $(0°, 0)$
Maximum at $(90°, \frac{1}{3})$
Minimum at $(270°, -\frac{1}{3})$

c The graph of $y = \sin\frac{1}{3}\theta$ is the graph of $y = \sin\theta$ stretched by a scale factor 3 in the θ direction.

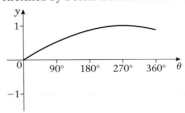

Only meets axis at origin
Maximum at $(270°, 1)$

d The graph of $y = \tan(\theta - 45°)$ is the graph of $\tan\theta$ translated by $45°$ to the right.

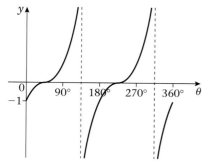

Meets θ-axis at $(45°, 0)$, $(225°, 0)$
Meets y-axis at $(0°, -1)$
[Asymptotes at $\theta = 135°$ and $\theta = 315°$]

4 a This is the graph of $y = \sin\theta°$ stretched by scale factor -2 in the y-direction (i.e. reflected in the θ-axis and scaled by 2 in the y direction).

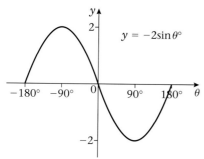

Meets θ-axis at $(-180°, 0)$, $(0, 0)$, $(180, 0)$
Maximum at $(-90°, 2)$
Minimum at $(90°, -2)$.

b This is the graph of $y = \tan\theta°$ translated by $180°$ to the left.

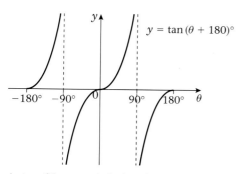

As $\tan\theta°$ has a period of $180°$
$\tan(\theta + 180)° = \tan\theta°$
Meets θ-axis at $(-180°, 0)$, $(0, 0)$, $(180°, 0)$
Meets y-axis at $(0, 0)$

c This is the graph of $y = \cos\theta°$ stretched by scale factor $\frac{1}{4}$ horizontally.

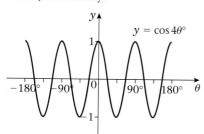

Meets θ-axis at $(-157\frac{1}{2}°, 0)$, $(-112\frac{1}{2}°, 0)$, $(-67\frac{1}{2}°, 0)$, $(-22\frac{1}{2}°, 0)$, $(22\frac{1}{2}°, 0)$, $(67\frac{1}{2}°, 0)$, $(112\frac{1}{2}°, 0)$, $(157\frac{1}{2}°, 0)$
Meets y-axis at $(0, 1)$
Maxima at $(-90°, 1)$, $(0, 1)$, $(90°, 1)$, $(180°, 1)$
Minima at $(-135°, -1)$, $(-45°, -1)$, $(45°, -1)$, $(135°, -1)$

d This is the graph of $y = \sin\theta°$ reflected in the y-axis. (This is the same as $y = -\sin\theta°$.)

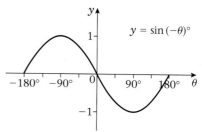

Meets θ-axis at $(-180°, 0)$, $(0°, 0)$, $(180°, 0)$
Maximum at $(-90°, 1)$
Minimum at $(90°, -1)$

5 a This is the graph of $y = \sin\theta$ stretched by scale factor 2 horizontally.
Period $= 4\pi$

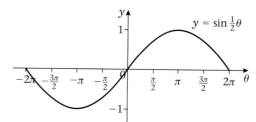

b This is the graph of $y = \cos\theta$ stretched by scale factor $-\frac{1}{2}$ vertically (equivalent to reflecting in θ-axis and stretching vertically by $+\frac{1}{2}$).
Period $= 2\pi$

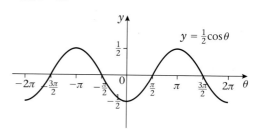

c This is the graph of $y = \tan\theta$ translated by $\frac{\pi}{2}$ to the right.

Period $= \pi$

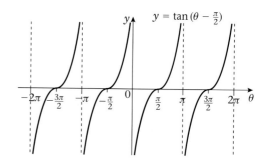

d This is the graph of $y = \tan \theta$ stretched by scale factor $\frac{1}{2}$ horizontally. Period $= \dfrac{\pi}{2}$

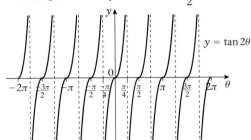

6 a i $y = \cos(-\theta)$ is a reflection of $y = \cos\theta$ in the y-axis, which is the same curve, so $\cos\theta = \cos(-\theta)$.

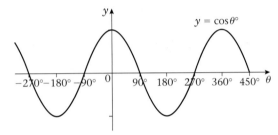

ii $y = \sin(-\theta)$ is a reflection of $y = \sin\theta$ in the y-axis.

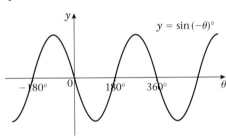

$y = -\sin(-\theta)$ is a reflection of $y = \sin(-\theta)$ in the θ-axis, which is the graph of $y = \sin\theta$, so $-\sin(-\theta) = \sin\theta$.

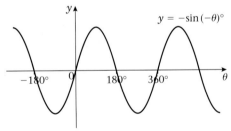

iii $y = \sin(\theta - 90°)$ is the graph of $y = \sin\theta$ translated by 90° to the right, which is the graph of $y = -\cos\theta$, so $\sin(\theta - 90°) = -\cos\theta$.

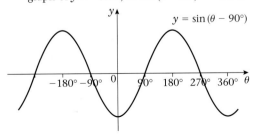

b $\sin(90° - \theta) = -\sin\{-(90° - \theta)\} = -\sin(\theta - 90°)$
using (a)(ii)
$= -(-\cos\theta)$ using (a)(iii)
$= \cos\theta$

c Using (a)(i) $\cos(90° - \theta) = \cos\{-(90° - \theta)\}$
$= \cos(\theta - 90°)$,
but $\cos(\theta - 90°) = \sin\theta$, so $\cos(90° - \theta) = \sin\theta$

Mixed exercise 8G

1 a $-\cos 57°$ **b** $-\sin 48°$ **c** $+\tan 10°$

d $+\sin 0.84^c$ **e** $+\cos\left(\dfrac{\pi}{15}\right)$

2 a -1 **b** $-\dfrac{\sqrt{2}}{2}$ **c** -1 **d** $+\sqrt{3}$ **e** -1

f $+\dfrac{\sqrt{3}}{2}$ **g** $-\dfrac{\sqrt{3}}{2}$ **h** $-\frac{1}{2}$ **i** 0 **j** $+\frac{1}{2}$

3 a A stretch of scale factor 2 in the x direction.
b A translation of $+3$ in the y direction.
c A reflection in the x-axis.
d A translation of $+20$ in the x direction (i.e. 20 to the right).

4 a

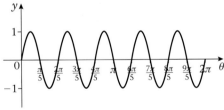

b There are no solutions of $\tan\left(x - \dfrac{\pi}{4}\right) + 2\cos x = 0$ in the interval $0 \leq x \leq \pi$, since $y = \tan\left(x - \dfrac{\pi}{4}\right)$ and $y = -2\cos x$ do not intersect in the interval.

5 a 300 **b** $(30, 1)$ **c** 60 **d** $\dfrac{\sqrt{3}}{2}$

6 a The graph is that of $y = \sin x$ stretched in the x direction. Each 'half-wave' has interval $\dfrac{\pi}{5}$.

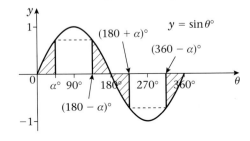

b $\dfrac{2\pi}{5}$ **c** 5

7 a The four shaded regions are congruent.

b $\sin \alpha°$ and $\sin(180 - \alpha)°$ have the same y value
(call it k)
so $\sin \alpha° = \sin(180 - \alpha)°$
$\sin(180 - \alpha)°$ and $\sin(360 - \alpha)°$ have the same y
value, (which will be $-k$)
so $\sin \alpha° = \sin(180 - \alpha)°$
$= -\sin(180 + \alpha)°$
$= -\sin(360 - \alpha)°$

8 a

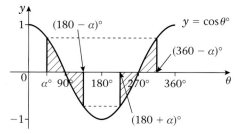

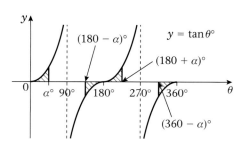

b i From the graph of $y = \cos \theta°$, which shows
four congruent shaded regions, if the y value
at $\alpha°$ is k, then y at $(180 - \alpha)°$ is $-k$, y at
$(180 - \alpha)° = -k$ and y at $(360 - \alpha)° = +k$
so $\cos \alpha° = -\cos(180 - \alpha)°$
$= -\cos(180 + \alpha)°$
$= \cos(360 - \alpha)°$

ii From the graph of $y = \tan \theta°$, if the y value at
$\alpha°$ is k, then at $(180 - \alpha)°$ it is $-k$, at $(180 + \alpha)°$
it is $+k$ and at $(360 - \alpha)°$ it is $-k$,
so $\tan \alpha° = -\tan(180 - \alpha)°$
$= +\tan(180 + \alpha)°$
$= -\tan(360 - \alpha)°$

Exercise 9A

1 a $x > -\frac{4}{3}$ **b** $x < \frac{2}{3}$
c $x < -2$ **d** $x < 2, x > 3$
e $x \in \mathbb{R}, x \neq 1$ **f** $x \in \mathbb{R}$
g $x > 0$ **h** $x > 6$
2 a $x < 4.5$ **b** $x > 2.5$
c $x > -1$ **d** $-1 < x < 2$
e $-3 < x < 3$ **f** $-5 < x < 5$
g $0 < x < 9$ **h** $-2 < x < 0$

Exercise 9B

1 a -28
b -17
c $-\frac{1}{5}$
2 a 10
b 4
c 12.25

3 a $(-\frac{3}{4}, -\frac{9}{4})$ minimum
b $(\frac{1}{2}, 9\frac{1}{4})$ maximum
c $(-\frac{1}{3}, 1\frac{5}{27})$ maximum, $(1, 0)$ minimum
d $(3, -18)$ minimum, $(-\frac{1}{3}, \frac{14}{27})$ maximum
e $(1, 2)$ minimum, $(-1, -2)$ maximum
f $(3, 27)$ minimum
g $(\frac{9}{4}, -\frac{9}{4})$ minimum
h $(2, -4\sqrt{2})$ minimum
i $(\sqrt{6}, -36)$ minimum, $(-\sqrt{6}, -36)$ minimum,
$(0, 0)$ maximum

4 a

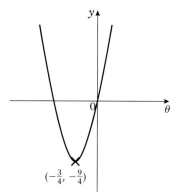

b

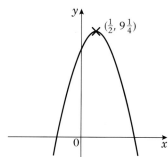

c

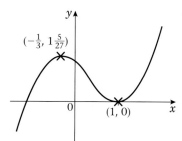

d

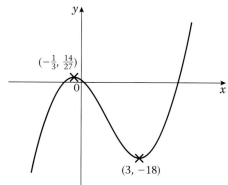

5 (1, 1) inflexion (gradient is positive either side of point)

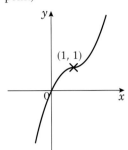

6 Maximum value is 27; $f(x) \leqslant 27$

Exercise 9C

1 $20\,\text{m} \times 40\,\text{m}$; $800\,\text{m}^2$

2 $200\pi\,\text{cm}^2$

3 $40\,\text{cm}$

4 $\dfrac{800}{4+\pi}\,\text{cm}^2$

5 $27\,216\,\text{mm}^2$

Mixed exercise 9D

1 a $x = 4,\ y = 20$

b $\dfrac{d^2y}{dx^2} = \dfrac{15}{8} > 0$ ∴ minimum

2 $(1, -11)$ and $(\tfrac{7}{3}, -12\tfrac{5}{27})$

3 a $7\tfrac{31}{32}$

b $\dfrac{x^3}{3} - 2x - \dfrac{1}{x} - 2\tfrac{2}{3}$

c $f'(x) = \left(x - \dfrac{1}{x}\right)^2 > 0$ for all values of x.

4 $(1, 4)$

5 a $y = 1 - \dfrac{x}{2} - \dfrac{\pi x}{4}$ **c** $\dfrac{2}{4+\pi}\,\text{m}^2$ $(0.280\,\text{m}^2)$

6 b $t = 2$ **c** $\sqrt{101} = 10.0$ (3 sf)

7 b $\dfrac{10}{3}$ **c** $\dfrac{d^2V}{dx^2} < 0$ ∴ maximum

d $\dfrac{2300\pi}{27}$ **e** $22\tfrac{2}{9}\%$

8 c $x = 20\sqrt{2},\ S = 1200\,\text{m}^2$

d $\dfrac{d^2S}{dx^2} > 0$

9 a $\dfrac{250}{x^2} - 2x$

b $(5, 125)$

10 b $x = \pm 2\sqrt{2}$, or $x = 0$

c $OP = 3$; $f''(x) > 0$ so minimum when $x = \pm 2\sqrt{2}$ (maximum when $x = 0$)

11 b A is $(-1, 0)$; B is $(\tfrac{5}{3}, 9\tfrac{13}{27})$

Exercise 10A

1 a $\sin^2 \dfrac{\theta}{2}$ **b** 5 **c** $-\cos^2 A$

d $\cos \theta$ **e** $\tan x^0$ **f** $\tan 3A$

g 4 **h** $\sin^2 \theta$ **i** 1

2 $1\tfrac{1}{2}$

3 $\tan x = 3 \tan y$

4 a $1 - \sin^2 \theta$ **b** $\dfrac{\sin^2 \theta}{1 - \sin^2 \theta}$ **c** $\sin \theta$

d $\dfrac{1 - \sin^2 \theta}{\sin \theta}$ **e** $1 - 2\sin^2 \theta$

5 (One outline example of a proof is given)

a $\text{LHS} \equiv \sin^2 \theta + \cos^2 \theta + 2 \sin \theta \cos \theta$
$\equiv 1 + 2 \sin \theta \cos \theta$
$= \text{RHS}$

b $\text{LHS} \equiv \dfrac{1 - \cos^2 \theta}{\cos \theta} \equiv \dfrac{\sin^2 \theta}{\cos \theta} \equiv \sin \theta \times \dfrac{\sin \theta}{\cos \theta}$
$\equiv \sin \theta \tan \theta = \text{RHS}$

c $\text{LHS} \equiv \dfrac{\sin x^°}{\cos x^°} + \dfrac{\cos x^°}{\sin x^°} \equiv \dfrac{\sin^2 x^° + \cos^2 x^°}{\sin x^° \cos x^°}$
$\equiv \dfrac{1}{\sin x^° \cos x^°} = \text{RHS}$

d $\text{LHS} \equiv \cos^2 A - (1 - \cos^2 A) \equiv 2 \cos^2 A - 1$
$\equiv 2(1 - \sin^2 A) - 1 \equiv 1 - 2\sin^2 A = \text{RHS}$

e $\text{LHS} \equiv (4\sin^2 \theta - 4\sin \theta \cos \theta + \cos^2 \theta)$
$+ (\sin^2 \theta + 4\sin \theta \cos \theta + 4\cos^2 \theta)$
$\equiv 5(\sin^2 \theta + \cos^2 \theta) = 5 = \text{RHS}$

f $\text{LHS} \equiv 2 - (\sin^2 \theta - 2\sin \theta \cos \theta + \cos^2 \theta)$
$\equiv 2(\sin^2 \theta + \cos^2 \theta)$
$- (\sin^2 \theta - 2\sin \theta \cos \theta + \cos^2 \theta)$
$\equiv \sin^2 \theta + 2\sin \theta \cos \theta + \cos^2 \theta$
$\equiv (\sin \theta + \cos \theta)^2 = \text{RHS}$

g $\text{LHS} \equiv \sin^2 x(1 - \sin^2 y) - (1 - \sin^2 x) \sin^2 y$
$\equiv \sin^2 x - \sin^2 y = \text{RHS}$

6 a $\sin \theta = \dfrac{5}{13},\ \cos \theta = \dfrac{12}{13}$

b $\sin \theta = \dfrac{4}{5},\ \tan \theta = -\dfrac{4}{3}$

c $\cos \theta = \dfrac{24}{25},\ \tan \theta = -\dfrac{7}{24}$

7 a $-\dfrac{\sqrt{5}}{3}$ **b** $-\dfrac{2\sqrt{5}}{5}$

8 a $-\dfrac{\sqrt{3}}{2}$ **b** $\dfrac{1}{2}$

9 a $-\dfrac{\sqrt{7}}{4}$ **b** $-\dfrac{\sqrt{7}}{3}$

10 a $x^2 + y^2 = 1$

b $4x^2 + y^2 = 4$ $\left(\text{or } x^2 + \dfrac{y^2}{4} = 1\right)$

c $x^2 + y = 1$

d $x^2 = y^2(1 - x^2)$ $\left(\text{or } x^2 + \dfrac{x^2}{y^2} = 1\right)$

e $x^2 + y^2 = 2$ $\left\{\text{or } \dfrac{(x+y)^2}{4} + \dfrac{(x-y)^2}{4} = 1\right\}$

Exercise 10B

1 a $270°$ **b** $60°, 240°$ **c** $60°, 300°$

d $15°, 165°$ **e** $140°, 220°$ **f** $135°, 315°$

g $90°, 270°$ **h** $230°, 310°$ **i** $45.6°, 134.4°$

j $135°, 225°$ **k** $30°, 210°$ **l** $135°, 315°$

m $131.8°, 228.2°$

n $90°, 126.9°, 233.1°$

o $180°, 199.5°, 340.5°, 360°$

2 a −120, −60, 240, 300   **b** −171, −8.63

c −144, 144   **d** −327, −32.9

e 150, 330, 510, 690   **f** 251, 431

3 a $-\pi, 0, \pi, 2\pi$ **b** $-\dfrac{4\pi}{3}, -\dfrac{2\pi}{3}, \dfrac{2\pi}{3}$

c $\dfrac{\pi}{4}, \dfrac{3\pi}{4}$ **d** $\pi, 2\pi$

e −0.14, 3.00, 6.14   **f** 0.59, 3.73

Exercise 10C

1 a 0°, 45°, 90°, 135°, 180°, 225°, 270°, 315°, 360°

b 60°, 180°, 300°

c $22\frac{1}{2}°, 112\frac{1}{2}°, 202\frac{1}{2}°, 292\frac{1}{2}°$

d 30°, 150°, 210°, 330°

e 300°

f 225°, 315°

g 90°, 270°

h 50°, 170°

i 165°, 345°

j 65°, 245°

2 a 250°, 310°

b 16.9°, 123°

c −77.3°, −17.3°, 42.7°, 103°, 163°

d −42.1°, −2.88°, 47.9°, 87.1°

3 a $-\dfrac{7\pi}{12}, -\dfrac{\pi}{12}$

b −0.986, 0.786

c $0, \dfrac{\pi}{2}, \pi, \dfrac{3\pi}{2}, 2\pi$

d 1.48, 5.85

Exercise 10D

1 a 60°, 120°, 240°, 300°

b 45°, 135°, 225°, 315°

c 0°, 180°, 199°, 341°, 360°

d 77.0°, 113°, 257°, 293°

e 60°, 300°

f 204°, 336°

g 30°, 60°, 120°, 150°, 210°, 240°, 300°, 330°

h 0°, 75.5°, 180°, 284°, 360°

i 270°

j 0°, 90°, 180°, 270°, 360°

k 0°, 18.4°, 180°, 198°, 360°

l 90°, 104°, 256°, 270°

m 72.0°, 144°, 216°, 288°

n 0°, 60°, 300°, 360°

o 194°, 270°, 346°

p 0°, 360°

2 a ±45°, ±135°   **b** −180°, −117°, 0°, 63.4°, 180°

c ±114°   **d** −127°, −73.4°, 53.4°, 107°

e ±180°, ±60°   **f** ±41.8°, ±138°

g 38.2°, 142°   **h** ±106°

3 a $\dfrac{\pi}{2}, \dfrac{3\pi}{2}$ **b** $\dfrac{5\pi}{12}, \dfrac{11\pi}{12}, \dfrac{17\pi}{12}, \dfrac{23\pi}{12}$

c 0, 0.983, π, 4.12, 2π **d** 0, 2.03, π, 5.18, 2π

e 0.841, $\dfrac{2\pi}{3}, \dfrac{4\pi}{3}$, 5.44   **f** 4.01, 5.41

g 1.11, 2.68, 4.25, 5.82

Mixed exercise 10E

1 Using $\sin^2 A = 1 - \cos^2 A$, $\sin^2 A = 1 - \left(-\sqrt{\dfrac{7}{11}}\right)^2 = \dfrac{4}{11}$.

Since angle A is obtuse, it is in the second quadrant and sin is positive, so $\sin A = +\dfrac{2}{\sqrt{11}}$.

Then $\tan A = \dfrac{\sin A}{\cos A} = \dfrac{2}{\sqrt{11}} \times \left(-\sqrt{\dfrac{11}{7}}\right) = -\dfrac{2}{\sqrt{7}} = -\dfrac{2}{7}\sqrt{7}$.

2 a $-\dfrac{\sqrt{21}}{5}$ **b** $-\dfrac{2}{5}$

3 a (−240, 0), (−60, 0), (120, 0), (300, 0)

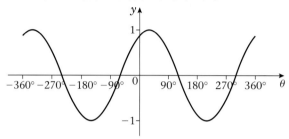

b (−270, 0), (−30, 0), (90, 0), (330, 0)

4 a $\cos^2 \theta - \sin^2 \theta$ **b** $\sin^4 3\theta$ **c** 1

5 a 1   **b** $\tan y = \dfrac{4 + \tan x}{2\tan x - 3}$

6 a LHS $\equiv (1 + 2\sin\theta + \sin^2\theta) + \cos^2\theta$
$\equiv 1 + 2\sin\theta + 1$
$\equiv 2 + 2\sin\theta$
$\equiv 2(1 + \sin\theta) =$ RHS

b LHS $\equiv \cos^4\theta + \sin^2\theta$
$\equiv (1 - \sin^2\theta)^2 + \sin^2\theta$
$\equiv 1 - 2\sin^2\theta + \sin^4\theta + \sin^2\theta$
$\equiv (1 - \sin^2\theta) + \sin^4\theta$
$\equiv \cos^2\theta + \sin^4\theta =$ RHS

7 a No solutions: $-1 \leqslant \sin\theta \leqslant 1$

b 2 solutions: $\tan\theta = -1$ has 2 solutions in the interval.

c No solutions: $2\sin\theta + 3\cos\theta > -5$ so $2\sin\theta + 3\cos\theta + 6$ can never be equal to 0.

d No solutions: $\tan^2\theta = -1$ has no real solutions.

8 a $(4x - y)(y + 1)$

b 14.0°, 180°, 194°

9 a $3\cos 3\theta$

b 16.1, 104, 136, 224, 256, 344

10 0.73, 2.41, 4.71

11 a $2\sin 2\theta = \cos 2\theta \Rightarrow \dfrac{2\sin 2\theta}{\cos 2\theta} = 1$
$\Rightarrow 2\tan 2\theta = 1 \Rightarrow \tan 2\theta = 0.5$

b 13.3, 103.3, 193.3, 283.3

12 a 225, 345

b 22.2, 67.8, 202.2, 247.8

13 a (0, 1)

b $\dfrac{17\pi}{24}, \dfrac{23\pi}{24}, \dfrac{41\pi}{24}, \dfrac{47\pi}{24}$

14 a $\dfrac{11\pi}{12}, \dfrac{23\pi}{12}$

b $\dfrac{2\pi}{3}, \dfrac{5\pi}{6}, \dfrac{5\pi}{3}, \dfrac{11\pi}{6}$

15 30°, 150°, 210°

16 36°, 84°, 156°

17 0°, 131.8°, 228.2°

18 a 90   **b** $x = 120$ or 300

19 a 60, 150, 240, 330
  b i (340, 0)   **ii** $p = \frac{3}{2}$, $q = 60$
20 a 30
  b At C (45, 0) the curve crosses the positive x-axis for the first time, so $1 + 2\sin(45p° + 30°) = 0$. This gives $(45p° + 30°) = 210°$.
  c B (15, 3), D (75, 0)

Exercise 11A

1 a $5\frac{1}{4}$ **b** 10   **c** $11\frac{5}{6}$ **d** $8\frac{1}{2}$ **e** $60\frac{1}{2}$
2 a $16\frac{2}{3}$ **b** $6\frac{1}{2}$ **c** $46\frac{1}{2}$ **d** $\frac{11}{14}$ **e** $2\frac{1}{2}$

Exercise 11B

1 a 8   **b** $9\frac{3}{4}$ **c** $19\frac{2}{3}$ **d** 21   **e** $8\frac{5}{12}$
2 4
3 6
4 $10\frac{2}{3}$
5 $21\frac{1}{3}$
6 $1\frac{1}{3}$

Exercise 11C

1 $1\frac{1}{3}$
2 $20\frac{5}{6}$
3 $40\frac{1}{2}$
4 $1\frac{1}{3}$
5 $21\frac{1}{12}$

Exercise 11D

1 a $A(-2, 6)$, $B(2, 6)$ **b** $10\frac{2}{3}$
2 a $A(1, 3)$, $B(3, 3)$ **b** $1\frac{1}{3}$
3 $6\frac{2}{3}$
4 4.5
5 a (2, 12)   **b** $13\frac{1}{3}$
6 a $20\frac{5}{6}$ **b** $17\frac{1}{6}$
7 c $y = x - 4$ **d** $8\frac{3}{5}$
8 $3\frac{3}{8}$
9 b 7.2
10 a $21\frac{1}{3}$ **b** $2\frac{5}{9}$

Exercise 11E

1 0.473
2 2.33
3 3.28
4 a 4.34   **b** Overestimate; curve bends 'inwards'
5 1.82
6 a 1.11   **b** Overestimate; curve bends 'inwards'
7 a

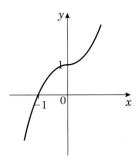

b 2
c 2
d Same; the trapezium rule gives an underestimate of the area between $x = -1$ and $x = 0$, and an overestimate between $x = 0$ and $x = 1$, and these cancel out.
8 2.87
9 a 1.59
  b Underestimate; curve bends 'outwards'
10 a i 1.8195   **ii** 1.8489
  b $\frac{4\sqrt{2}}{3}$ **i** 3.51%   **ii** 1.95%

Mixed exercise 11F

1 a $-1, 3$ **b** $10\frac{2}{3}$
2 a $q = 18$ **b** $73\frac{1}{4}$
3 a $\frac{4}{3}\sqrt{3}$ **b** $\frac{168}{5}\sqrt{2} - \frac{162}{5}$
4 a $5x - 8x^{\frac{1}{2}} - \frac{2}{3}x^{\frac{3}{2}} + c$ **b** $2\frac{1}{3}$
5 a (3, 0)   **b** (1, 4)   **c** $6\frac{3}{4}$
6 a $A = 6, B = 9$
  b $\frac{3}{5}x^{\frac{5}{3}} + \frac{9}{2}x^{\frac{4}{3}} + 9x + c$
  c $149\frac{1}{10}$
7 a $\frac{3}{2}x^{-\frac{1}{2}} + 2x^{-\frac{3}{2}}$
  b $2x^{\frac{3}{2}} - 8x^{\frac{1}{2}} + c$
  c $6 - 2\sqrt{3}$
8 a $2y + x = 0$ **b** $y = -1\frac{1}{4}$ **c** $2\frac{29}{48}$
9 b (4, 16)   **c** 133 (3 sf)
10 a (6, 12)   **b** $13\frac{1}{3}$
11 a $A(1, 0)$, $B(5, 0)$, $C(6, 5)$ **b** $10\frac{1}{6}$
12 a 1
  b $y = x + 1$
  c $A(-1, 0)$, $B(5, 6)$ **d** $50\frac{5}{6}$
  e $33\frac{1}{3}$
13 a $q = -2$ **b** $C(6, 17)$ **c** $1\frac{1}{3}$

Examination style paper (C2)

1 a 17 cm   **b** 17.5 cm^2
2 a $3 + 2p$ **b** 32
3 a 7   **b** $(x - 3)(x + 2)(x + 1)$
4 a 21 840; 3640   **b** 6
5 a LHS $\equiv \dfrac{1 - \sin^2\theta}{\sin\theta(1 + \sin\theta)} \equiv \dfrac{(1 - \sin\theta)(1 + \sin\theta)}{\sin\theta(1 + \sin\theta)} \equiv$ RHS
  b 19, 161
6 a 5   **b** $2\sqrt{10}$
7 b £225   **c** 6
8 a 9.084 25   **b** $6\sqrt{3} - \frac{4}{3}$ **c** 0.3
9 a $14x^{\frac{4}{3}} - 14x$
  b $\frac{56}{3}x^{\frac{1}{3}} - 14$
  c (0, 4) maximum, (1, 3) minimum

Index